HEMATOPOIETIC AND LYMPHOID TISSUE IN CULTURE

STUDIES IN SOVIET SCIENCE

LIFE SCIENCES

1973

MOTILE MUSCLE AND CELL MODELS
  N. I. Arronet
PATHOLOGICAL EFFECTS OF RADIO WAVES
  M. S. Tolgskaya and Z. V. Gordon
CENTRAL REGULATION OF THE PITUITARY-ADRENAL COMPLEX
  E. V. Naumenko

1974

SULFHYDRYL AND DISULFIDE GROUPS OF PROTEINS
  Yu. M. Torchinskii
MECHANISMS OF GENETIC RECOMBINATION
  V. V. Kushev

1975

THYROID HORMONES: Biosynthesis, Physiological Effects, and
  Mechanisms of Action
  *Ya. Kh. Turakulov, A. I. Gagel'gans, N. S. Salakhova, A. K. Mirakhmedov,
  L. M. Gol'ber, V. I. Kandror, and G. A. Gaidina*

1977

THE EVOLUTIONARY ECOLOGY OF ANIMALS
  S. S. Shvarts
HEMATOPOIETIC AND LYMPHOID TISSUE IN CULTURE
  E. A. Luriya
STRUCTURE AND BIOSYNTHESIS OF ANTIBODIES
  R. S. Nezlin
PROTEIN METABOLISM OF THE BRAIN
  A. V. Palladin, Ya. V. Belik, and N. M. Polyakova

A Continuation Order Plan is available for this series. A continuation order will bring
delivery of each new volume immediately upon publication. Volumes are billed only upon
actual shipment. For further information please contact the publisher.

STUDIES IN SOVIET SCIENCE

HEMATOPOIETIC AND LYMPHOID TISSUE IN CULTURE

E. A. Luriya

Academy of Medical Sciences of the USSR
Moscow, USSR

Translated from Russian by
Basil Haigh

With a foreword by
Robert Auerbach
University of Wisconsin
Madison, Wisconsin

CONSULTANTS BUREAU • NEW YORK AND LONDON

Library of Congress Cataloging in Publication Data

Luriĩa, Elena Aleksandrovna.
  Hematopoietic and lymphoid tissue in culture.

  (Studies in Soviet science)
  Translation of Krovetvornaĩa i limfoidnaĩa tkan' v kul' turakh.
  Bibliography: p.
  Includes index.
  1. Lymphoid tissue. 2. Hematopoietic system. 3. Immunocompetent cells. 4. Cell
culture. I. Title. II. Series. [DNLM: 1. Hematopoietic system—Cytology. 2. Lympho-
cytes. 3. Lymphoid tissue. WH700 L967K]
QP115.L8713   599'.08'21   76-55703
ISBN 978-1-4684-1631-2   ISBN 978-1-4684-1629-9 (eBook)
DOI 10.1007/978-1-4684-1629-9

Elena Aleksandrovna Luriya was born in 1938 and is a graduate of the Biological
Faculty of Moscow University. Her main field of research is the cultivation of hematopoietic and lymphoid tissue in diffusion chambers and in organ and monolayer
cultures in vitro. At present her post is Research Scientist in the laboratory of immunomorphology of the N.F. Gamaleya Institute of Epidemiology and Microbiology,
Academy of Medical Sciences of the USSR, Moscow.

The original Russian text, published by Meditsina Press in Moscow in 1972, has been
corrected by the author for the present edition. This translation is published under an
agreement with the Copyright Agency of the USSR (VAAP).

КРОВЕТВОРНАЯ И ЛИМФОИДНАЯ ТКАНЬ В КУЛЬТУРАХ

Е. А. ЛУРИЯ

KROVETVORNAYA I LIMFOIDNAYA TKAN' V KUL'TURAKH

E.A. Luriya

©1977 Consultants Bureau, New York
Softcover reprint of the hardcover 1st edition 1977
A Division of Plenum Publishing Corporation
227 West 17th Street, New York, N.Y. 10011

Foreword

This translation from Russian into English of the monograph *Hematopoietic and Lymphoid Tissue in Culture* by E. A. Luriya provides us for the first time with a comprehensive review of the massive amount of work carried out in the USSR in this important area of biological research. Historically, the pioneering tissue culture studies of Maksimov, Popov, Chasovnikov, and their colleagues working in Russia in the early part of this century provided much of the impetus for subsequent studies on lymphopoiesis and hematopoiesis *in vitro* and much of the basic information used subsequently to achieve immune responses *in vitro* and to develop the methodology for cloning hematopoietic stem cells.

Particularly welcome in this translation is the addition of the Appendix, in which Drs. Luriya and Fridenshtein describe in some detail the methodology that has been developed in their laboratories for the study of a variety of cell types in cell and tissue culture. For many of us, this description provides information that had previously not been accessible to us: porosity of millipore filters in relation to growth properties; colony-forming cell adhesiveness; proliferation kinetics of stromal cells; differential effectiveness of various media used for culture of liver cells; details of washing procedures for filter substrates used in culture.

Similarly, while most of us have been well aware of the contributions to the literature made by research work carried out by Drs. Fridenshtein and Luriya, investigators such as Kolesnikova, Chailakhyan, Chertkov, Latsinik, and Keilis-Borok, whose names appear as senior authors in studies of obvious significance in the field, are virtually ignored outside the USSR because of the restriction of their published work to the Soviet literature.

In reading this translation, it should be kept in mind that this is a monograph intended primarily to communicate to the scientific community the work

v

carried out in the USSR dealing with the subject of *in vitro* analysis of the blood-forming system. The literature review is largely intended to provide a historical perspective. And, with the exception of the Appendix, the manuscript was completed and published in 1972, though the author has revised the text and added recent developments for this English edition.

I for one am delighted that this translation with its Appendix is being published. I consider it to be an important addition to the literature for those of us unable to read Russian, and valuable as a resource for subsequent reviews so that the studies described in this monograph can be cited and placed in their proper historical and scientific perspective.

Robert Auerbach

Foreword to the English Edition

Fifty years ago, Aleksandr Aleksandrovich Maksimov carried out his classic experiments on explanation of hematopoietic and lymphoid tissue in plasma clots, and in so doing laid the foundations of future research into the cultivation of hematopoietic and immunocompetent cells. Although the analysis of histogenesis of cell transformations is just as important today as ever, it is only one of the many applications of tissue culture work. *In vitro* systems are now used extensively to study immunity and the regulation of hematopoiesis, and also as a method of isolating long-life cell lines, capable of differentiating in various ways, from hematopoietic tissue.

Research with hematopoietic tissue cultures is being pursued intensively in many laboratories. Interest in this field has grown in particular in connection with the development of methods of cloning hematopoietic cells *in vitro*. Special symposia and working conferences held during recent years have shown that the pool of workers studying hematopoiesis *in vitro* is large and continues to grow.

For this English edition, the text has been extensively revised and updated, and considerable new material has been added, in particular the Appendix.

The main interest of Luriya's book for Western readers will probably be the full account of results obtained in the Gamaleya Institute in cloning stromal mechanocytes and in hematopoietic tissue organ cultures.

A. Ya. Fridenshtein
The Gamaleya Institute of Epidemiology and Microbiology
Academy of Medical Sciences of the USSR, Moscow

Preface

This monograph discusses the *in vitro* culture of hematopoietic and lymphoid tissue. Results obtained by the use of various explanation methods, each allowing certain particular aspects of the histogenesis, differentiation, and function of hematopoietic and lymphoid tissues to be analyzed, are compared.

The main part of the book presents the results of explantation of lymphoid and hematopoietic tissue in organ and monolayer cultures obtained by the author and her collaborators at the Gamaleya Institute in Moscow. By the multiple organ culture method, proliferation and differentiation of lymphoid and myeloid cells was studied for several weeks; in organ cultures of embryonal mouse liver, a line of hematopoietic stem cells was maintained for over 8 weeks. Using monolayer cultures, a colony-assay method for cloning stromal precursors from different populations of hematopoietic and lymphoid cells was developed, and diploid strains of stromal mechanocytes were obtained from different hematopoietic organs.

The book presents both the results of recent investigations and a detailed account of the earlier studies carried out by classic methods. In the author's opinion, some of these methods of hematopoietic and lymphoid tissue cultivation that are unpopular nowadays may prove fruitful in future research.

Contents

Introduction 1
  The Study of Hematopoietic and Lymphoid Tissues in Tissue
    Culture 1
  The Development of Tissue Culture Methods 5

Chapter I. Hematopoietic Tissue *In Vitro* 13
  1. Hematopoietic Tissue in Plasma Clot Culture 13
  2. Monolayer Cultures of Hematopoietic Tissue 16
  3. Suspension Cultures of Hematopoietic Tissue 29
  4. Formation of Hematopoietic Foci in Agar Cultures 30
  5. Organ Cultures of Hematopoietic Tissue 35
    Embryonic Liver 35
    Bone Marrow 48
  6. A Comparative Assessment of the Various Methods of
     Hematopoietic Tissue Culture 54
    Duration of Hematopoiesis 54
    Myelopoiesis 55
    Erythropoiesis 58
  7. Factors That Influence Differentiation of Hematopoietic Tissue
     In Vitro 59

Chapter II. Possibilities of Differentiation of Cells Grown in
  Hematopoietic Tissue Cultures on Retransplantation *In Vivo* 63

Chapter III. Lymphoid Tissue *In Vitro* 79
  1. Lymphoid Tissue Explanted in a Plasma Clot 79
    Spleen 79

Lymph Nodes 80
Thymus 81
Cells of the Peripheral Blood and Lymph 82
2. Monolayer Cultures of Lymphoid Tissue and Peripheral
   Blood Cells 85
3. Suspension Cultures of Lymphoid Tissue and Peripheral
   Blood Cells 91
4. Organ Cultures of Lymphoid Tissue 93
   Thymus 94
   Lymph Nodes 99
5. Comparison of Various Methods of Lymphoid Tissue Culture 107
   The Behavior of Lymphocytes 107
   Cell Transformations 109
   The Process of Regeneration 110
   Formation of Structures Characteristic of Interaction Between
   Lymphocytes and Stromal Cells 111

Chapter IV. Immunological Functions of Lymphoid Tissue *In Vitro* 113
1. The Process of Antibody Formation 113
   Basic Mechanisms of Antibody Formation *In Vivo* and Aims of
   Research into This Process *In Vitro* 113
   Secondary Antibody Synthesis *In Vitro* 116
   Primary Antibody Synthesis *In Vitro* 120
2. The Delayed Hypersensitivity Reaction *In Vitro* 127
   Basic Mechanisms of the Delayed Hypersensitivity Reaction
   In Vivo and Aims of Research into This Process *In Vitro* 127
   Blast-Transformation, Recognition of Antigens, and the Cellular
   Immune Response *In Vitro* 128
   In Vitro Destruction of Target Cells by Lymphocytes from
   Preimmunized Donors 131
   Inhibition of Macrophage Migration 138
   The Simonsen Phenomenon 140
   Immunological Tolerance 141
   Conclusion 142

Chapter V. Cell Lines in Lymphoid and Hematopoietic Tissue 145
   Conclusion 149

Appendix. Recent Experience in Monolayer and Organ Cultures of
   Hematopoietic Tissue — E. A. Luriya and A. Ya. Fridenshtein 151
1. The Technique of Cloning Fibroblasts in Monolayer Cultures of
   Hematopoietic and Lymphoid Cells and Its Use to Study Stromal
   Precursor Cells 151

2. The Method of Multiple Organ Cultures of Hematopoietic and
   Lymphoid Tissue 158

References 165

Index 183

Introduction

The Study of Hematopoietic and Lymphoid Tissues in Tissue Culture

The culture of lymphoid and hematopoietic tissue *in vitro* is an important problem from both the theoretical and the practical points of view. Attempts to obtain cultures of these tissues in which their differentiation could continue have been undertaken repeatedly since 1910. It proved impossible to maintain hematopoiesis or lymphopoiesis for any length of time by explantation of hematopoietic and lymphoid tissues by classic experimental methods. Only recently, with the development of new tissue culture techniques and approaches, has it become possible to grow these tissues *in vitro* so that the constituent cells can continue to differentiate for a sufficient length of time. Although the problem of culturing hematopoietic and lymphoid tissue cannot be regarded as completely solved, the progress already made has provided the research worker with a valuable model with which to study lymphopoiesis and hematopoiesis and their regulation.

Lymphoid and hematopoietic tissues are constructed according to the repopulation principle: many of the cells comprising them can migrate into the bloodstream and can survive in the hematopoietic and lymphoid organs to which they are carried by the bloodstream (Metcalf and Moore, 1971).

Lymphoid and hematopoietic tissues share a common ancestral or stem cell, giving rise to lymphopoiesis under some conditions and to hematopoiesis under others. Subsequent differentiation of hematopoietic and lymphoid cells *in vivo* is under the influence of many different factors in the local environment of the cells, including the hormonal and antigenic background, the concentration of vitamins and many other biologically active substances, and the interaction with cells of their own and other types. We know how some of these factors act, but we can still only guess at the actions of others. In a complex system such as the living organism, it is difficult, or even impossible, to evaluate and assess the quantitative role of each individual factor acting on the histogenesis of hematopoietic and lymphoid tissue. By using a transplantation method in an isolated system *in vivo*, namely, diffusion chambers, in which the repopulation process can be eliminated from the experimental conditions, valuable results have already been obtained by the study of cell transformation in hematopoietic tissue. In many respects, however, the scope of the diffusion chamber method is limited.

In tissue culture conducted *in vitro*, repopulation is only one of the processes eliminated from the experimental conditions. A tissue can be kept in culture under standardized and strictly controlled conditions and a simplified system created in which the role of individual factors in histogenesis can be analyzed more systematically. For example, the composition of the nutrient medium can be varied by adding appropriate substances in known concentrations. Also, the time of addition and the period of exposure of the tissue to the active factors are more easily regulated in tissue culture experiments.

Hematopoietic tissue cultures provide an extremely suitable model for analysis of the histogenetic action and the "points of application" of substances such as erythropoietins and leukopoietins, and they provide new opportunities for the experimental study of factors determining the direction of differentiation of hematopoietic precursors. They can also be used successfully to analyze the action of ionizing radiation on hematopoietic and lymphoid tissues. The cell kinetics of these tissues, involving proliferation and differentiation, can be studied easily in culture.

Isolation of individual cell lines from hematopoietic and lymphoid tissues and the definition of various categories of precursor cells found in these tissues are facilitated by tissue culture methods. Recently developed methods of cloning hematopoietic tissue cells in culture are now being used in connection with research in all these important fields.

Many of the problems connected with the cellular mechanisms of immunity also require tissue culture methods for their study. For a long time, it was impossible to induce a primary immune response *in vitro*, but now practically all the manifestations of immunity can be successfully provoked *in vitro*. Many important aspects of immunological reactions still await explanation and require analysis by tissue culture methods. Problems of particular interest are the roles of interaction among cells and of individual tissue structures in the mechanism of the successive phases of the immune response.

Tissue culture techniques thus provide the facilities for the study of important problems concerning the structure, differentiation, and function of lymphoid and hematopoietic tissues, as well as an indispensable model with which to analyze the action of biologically active substances on cells participating in immunological reactions and in hematopoiesis.

Hematopoietic tissue culture can prove to be of importance in connection with practical problems in medicine. Cell suspensions of lymphoid and hematopoietic tissues can be transplanted successfully to repopulate the lymphopoietic and hematopoietic organs of the recipient, where they subsequently undergo normal proliferation and differentiation. For example, injection of a suspension of bone marrow cells can protect a recipient against the otherwise lethal effects of prior irradiation. Hematopoiesis in the irradiated recipient is restored by proliferation and differentiation of hematopoietic stem cells, which constitute an enduring, self-maintaining cell line. A characteristic feature distinguishing lymphoid and hematopoietic tissues is the ability of the tissue to perform its functions completely if transplanted as a suspension of individual cells. In this respect, lymphoid and hematopoietic tissues have the advantage over kidney, muscle, and other tissues, for if these tissues are transplanted as separate cells, the functions of the corresponding organs are not restored.

It has recently been shown that hematopoietic stem cells can persist and proliferate in hematopoietic tissue cultures. Because they can, the prospects for the maintenance of hematopoietic tissue in cultures suitable for transplantation are presently on a firm theoretical basis. Needless to say, if hematopoietic cells, whether or not they have been cultured, are to be transplanted, the problem of immunological attack against the recipient by the donor cells still remains to be solved. However, the capability of maintaining hematopoietic tissue in culture allows the possibility that appropriate manipulations may be brought to bear on the immune reactivity of the culture cells, which will make them more suitable for further transplantation.

The range of problems that tissue culture is now being called on to solve is extremely large. It is not surprising, therefore, that the methods of culture in current use differ, sometimes fundamentally, depending on their purpose.

Methods of culture of hematopoietic and lymphoid tissue can be classified in three groups, each with its own particular features:

1. Short-term cultures are intended for the isolation of hematopoietic cells from the donor, and they satisfy the conditions for maintaining metabolism of the cells and for the performance of certain functions not requiring prolonged proliferation or progressive, stage-by-stage, differentiation. Short-term cultures are used to assess metabolic activity (including that of nucleic acids and proteins) in hematopoietic and lymphoid cells, to study the reactions of these cells to antigens and mitogens, to determine the parameters of the mitotic cycle and the time of maturation of certain categories of hematopoietic cells, and to clone hematopoietic precursor cells.

2. Long-term culture (for weeks or months), during which intensive cell

proliferation takes place; under these conditions, the cell mass may grow considerably in size. By no means, however, can all the types of cells composing hematopoietic and lymphoid tissue be propagated in such cultures. Hematopoiesis and lymphopoiesis under these conditions usually cease rapidly, and only those cells that are generally identified with the stroma of these organs will continue to grow. Cultures of this type are used to study problems connected with the histogenesis of hematopoietic tissue, with the isolation of pure cell lines from them, and so on.

3. Prolonged cultures in which proliferation and differentiation of the major types of hematopoietic and lymphoid cells continue to occur over extended periods of time, i.e., cultures in which hematopoiesis and lymphopoiesis take place. This does not mean that the various types of hematopoietic cells that have already begun to differentiate mature completely; rather, the method permits the maintenance in culture of a line of stem cells which can proliferate and develop in the same directions as they would *in vivo*. However, although such cultures are of particular interest for the investigation of many of the problems outlined above, they have not yet been widely used because of the great difficulties that arise in attempts to culture hematopoietic and lymphoid tissues and maintain their specific differentiation *in vitro*. This particular problem has recently been tackled successfully by the use of organ culture methods.

The main purpose of this book is to examine problems connected with the use of long-term cultivation of lymphoid and hematopoietic tissues. These problems include the isolation of stromal cell lines from hematopoietic and lymphoid tissue and the study of the histogenetic properties and the capacity of differentiation of cells of these lines under different experimental conditions (including transplantation *in vivo*). Particular attention is paid to the maintenance of hematopoiesis and lymphopoiesis in organ cultures. Only as much consideration as is necessary to this main purpose is given to the results of short-term culture experiments and to immunological processes *in vitro*.

Although hematopoietic and lymphoid tissues in the living organism constitute in many respects a single system, these two types of tissues are separated to varying degrees in the hematopoietic organs of different animals. In particular, rodents—the tissues of which are most frequently used for explantation—are characterized by a "lymphoid" type of hematopoiesis, in which the number of lymphocytes in the peripheral blood and of lymphoid cells in the bone marrow is higher than in man. Thus, despite an evident basic unity of hematopoietic and lymphoid tissues, there are substantial functional and, what is particularly important for explantation, histogenetic differences between hematopoietic and lymphoid tissues. The problems connected with culture of each of these tissues are therefore best examined separately. Methods of explantation of hematopoietic and lymphoid tissue in current use seemingly have little in common. If we look at the history of tissue culture, however, we can see that one method has arisen from the other. In this way, it is easier to assess the possibilities and limitations of each method of tissue culture.

The Development of Tissue Culture Methods

It is generally considered that the experiments of Roux in 1885 marked the beginning of the tissue culture method. Roux placed pieces of the medullary plate of a chick embryo in warm salt solution and was able to maintain this tissue in a viable condition *in vitro* for several days. The true date of the birth of the tissue culture method, however, is 1907, when Harrison described an easily reproducible technique of tissue culture that not only preserved viable tissue but also permitted cell proliferation. Harrison explanted fragments taken from the region of the medullary tube of a frog embryo into a clot of frog's lymph. Under these conditions, the pieces of tissue survived for several weeks, and differentiation of nerve cells and the outgrowth of axons were observed. It is interesting to note that even though Harrison used nerve tissue for his experiments, subsequent experience for more than 60 years showed that this tissue is one of the most difficult to grow in culture; in nerve explants, it is

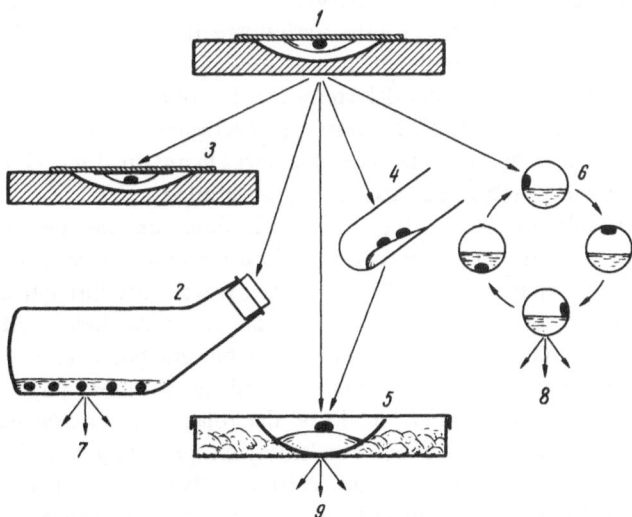

Fig. 1. The principal methods of tissue culture (1) hanging-drop culture in a plasma clot (Harrison, 1907); (2) culture in Carrel's flask in a lying drop (Carrel, 1923); (3) hanging-drop culture over a well with two cover slips (Maksimov, 1925); (4) culture on the surface of a plasma clot on the wall of a test tube (Strangeways and Fell, 1926); (5) culture on the surface of a plasma clot in a watch glass (Fell and Robinson, 1929); (6) culture in rolling tubes (Gey, 1933); (7) various methods of monolayer culture; (8) various methods of suspension culture; (9) various methods of organ culture.

chiefly the glial cells that proliferate, and special conditions of culture are needed to permit differentiation of the neuron. Nevertheless, Harrison's culture, in which the explant was placed in a drop of clotted lymph and was surrounded by air, proved to be a promising method of supporting the complex differentiation of nerve cells.

The method that Harrison employed for these early fragment cultures was a simple one. The piece of tissue for explantation is placed on a cover slip in a drop of lymph that is then allowed to set in order to form a gelatinous clot around the fragment. The cover slip with the clot attached is inverted and placed on a large hollow-ground slide and sealed with paraffin (Fig. 1). This technique, subsequently known as *hanging-drop culture,* is still in use today.

Whereas the honor for the discovery of the tissue culture method belongs to Harrison, most of the work on the development and improvement of the method and its introduction into laboratory practice on a wide scale was undertaken by Carrel and his co-workers. Carrel and Burrows (1910*a,b*) chose natural media to culture the tissues of warm-blooded animals; they used a clot of plasma instead of lymph and demonstrated that growth could be stimulated by the presence of chick embryo extract in the clot. The grafted fragment became surrounded by a "halo" consisting of cells that migrated from the explant. After a few days, the plasma clot became liquefied and growth was diminished. They also found that the explant must be transferred into fresh nutrient medium every 2 or 3 days in order for growth to continue in the cultures. The addition of chick embryo extract proved to be an essential condition for obtaining cultures capable of prolonged growth. In Carrel's laboratory, "everlasting" cultures were obtained from chick embryo heart tissue. This strain was maintained for 34 years. During this period, the cells remained capable of proliferation.

It is difficult to assess fully the work done on the development and maintenance of cultures in the period before the appearance of antibiotics, of ultraviolet irradiation, and of artificial media. It was only through the efforts of the eminent surgeon Carrel that the development of the tissue culture method and its introduction into biology and medicine became possible.

During the first years of existence of tissue cultures, physiological and biochemical investigations were made of the roles of individual factors of the nutrient medium: the effect of salt concentration, pH, temperature, and osmotic pressure. The metabolism of the cultured tissue was investigated, and physiological studies of the cells in the zone of growth were made.

The tissue culture method achieved wide popularity in the 1920s. Workers who began to use it included Krontovskii (1925), Maksimov (1922), Rumyantsev (1932), Fischer (1930), and Erdmann (1917). The work of each contributed to further advances in the technique and to a broadening of the range of problems studied by it.

Krontovskii (1925) used Gabrichevskii's dishes for explantation in a plasma clot, and in 1923, Carrel (1923) suggested special flasks for tissue culture. Carrel's flask with a flat bottom and tapering neck is a convenient vessel for growing tissues in a plasma clot for many months or even years. Its use simplifies the technique of working with tissue cultures, and many pieces of tissue can be grown in culture simultaneously in a large volume of medium.

In 1925, Maksimov (1925) suggested the double cover slip method. Essentially, this method has much in common with the hanging-drop method, but differs in that the cover slip with the fragment in the plasma clot is covered by another cover slip (or mica disk) of larger size, ensuring the sterility of the smaller cover slip. When the plasma clot liquefies, all that is necessary is to wash the small cover slip with the explant in saline solution, apply a drop of plasma and embryo extract, and resume cultivation.

A further improvement in the plasma culture method—the roller tube method—was introduced by Gey (1933). Tubes with fragments stuck to the walls by means of plasma clots, are placed in a special drum that revolves slowly, so that the liquid component of the medium (dilute embryo extract or serum) bathes the explant and is constantly mixed.

The technique of culture in a plasma clot has been applied to hematopoietic and lymphoid tissue in two modifications.

1. Small fragments of hematopoietic organs are mounted in a plasma clot consisting of heterologous or autologous plasma and tissue extract (most commonly embryonic). The clot is stuck to the surface of a test tube, to the bottom of a Carrel dish, or to a cover slip in a Maksimov dish. Liquid culture medium, which can be easily renewed during cultivation, can be added to the system without having to reseed the fragments, because hematopoietic tissue liquefies the plasma clot only slightly.

2. Blood cells or suspensions of cells prepared from hematopoietic organs are mounted in the plasma clot. Cultivation proceeds by the methods described above.

In the 1950s, as a result of advances in chemistry, the first synthetic media appeared. Their composition included amino acids, vitamins, and other biologically active substances. Meanwhile, methods of trypsinization of tissue were developed. This marked the beginning of a new period of evolution of tissue cultures: large numbers of cells could now be grown in monolayer and suspension cultures (Eagle, 1955; Moscona, 1952).

The tissue culture method has been widely adopted in laboratory practice, and has been used in many branches of experimental biology and medicine. Strains of cells on which viruses will grow have been obtained for use in virological research (Rapp and Melnick, 1966).

Tissue culture methods used in virology provide for intensive proliferation of cells and the production of large cell masses *in vitro*. As a rule, however, the cells in mass cultures (monolayer and suspension) lose their usual differentiation.

Monolayer and suspension cultures are widely used for explantation of hematopoietic and lymphoid tissue, usually with the following modifications:

1. Suspensions of cells from hematopoietic and lymphoid organs or blood cells are prepared in a synthetic culture medium with or without the addition of serum. Suspension cultures are frequently set up under special conditions to prevent adhesion of the cells to the surfaces of the culture vessels. This is done by shaking, by stirring, by blowing bubbles of gas through the medium (a method that in addition allows the medium to be kept at a given level of acidity and oxygen concentration), and also by special treatment of the surfaces of the culture vessels.

Lymphoid and hematopoietic cells can retain their viability in suspension cultures for several days, so that it is possible to study their metabolism (including incorporation of radioactive isotopes) and their response to immunological factors, mitogens, antimetabolites, etc.

2. Monolayer cultures. Cells of hematopoietic and lymphoid tissues, as well as blood cells, are placed in flat-bottomed vessels or in test tubes with cover slips, and cultured in synthetic medium with or without the addition of serum. Observations are made on the adherent cells spreading over the surface of the bottom of the vessel. In particular, proliferation of these cells, their movements and phagocytic activity, can be studied in monolayer cultures. As a rule, growth of histiocytes and fibroblasts without the maintenance of hematopoiesis is observed in such cultures for long periods of time (months).

Perhaps the most promising method for studying cell differentiation and morphogenesis *in vitro* is with the use of small fragments of tissue explanted in organ culture.

In 1926, Strangeways and Fell (1926) described a new method of cultivation on the surface of a plasma clot attached to the wall of a test tube. Later, Fell and Robinson (1929) introduced cultures in watch glasses. A watch glass 4 cm in diameter is placed in a petri dish, on the bottom of which is placed a pad of cotton wool liberally soaked with water. Chick plasma and embryo extract are poured drop by drop into the watch glass. After they are mixed, a clot is formed, and the explant is placed on its surface. Under these conditions of culture, Fell and co-workers observed the patterns of morphogenesis of the limb anlagen of the chick embryo, with the formation of joints and the development of cartilage and bone tissue.

Fell's investigations laid the foundations of a new branch of tissue culture—organ culture (Fig. 2), in which the explant is placed on the boundary with the gaseous phase. These conditions proved to be suitable for complex differentiation and tissue interaction *in vitro*. For a long time, Fell's method was the only method of organ culture, and not until many years later, in the 1950s, were other techniques of organ culture suggested and applied.

The main distinguishing feature of the organ culture method is that condi-

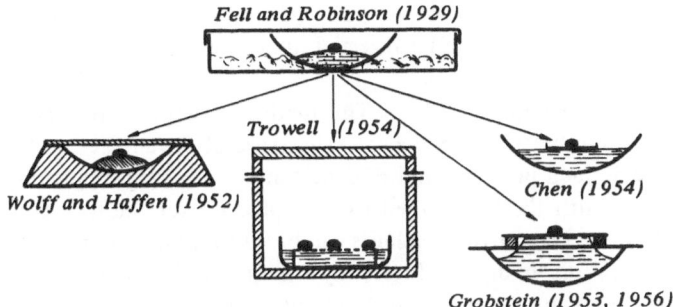

Fig. 2. Methods of organ culture.

tions are created for organ-specific differentiation with moderate growth of the tissue. In all modifications of the organ culture method, the tissue is placed on the surface of the nutrient medium. In this way, the explant is in contact with the gaseous phase, which may be air or a mixture of air or oxygen with carbon dioxide. Cultivation at the boundary between two phases is the basic condition for the organ culture method.* In organ cultures, whole organs or their fragments and cell suspensions can be explanted successfully on the boundary with the gaseous phase.

The methods of organ culture of explants of hematopoietic and lymphoid tissues can be classified as follows:

I. Culture on the Surface of a Plasma Clot in a Watch Glass. This method, developed by Fell and Robinson (1929), is the original method for all types of organ culture.

II. Culture on Semisolid Media Containing Agar. Instead of the plasma clot, a 0.3–0.5% solution of agar has been suggested (Spratt, 1947; Wolff and Haffen, 1952). Various nutrient components are added to the medium: a salt solution or synthetic medium, serum, embryonic extract, etc. Several formulas are used for nutrient media to be added to agar cultures. The substrate for growth of the explanted tissues is the surface of the agar gel. To prevent migration of the cells from the explant, Wolff (1960) suggested that it be covered with the vitelline membrane of a hen's egg.

III. Culture on the Surface of Liquid Media. Several methods, differing not only in the composition of their nutrient medium but also in the supporting substrate, on which the tissue is placed, have been suggested:

(1) Cultivation on a metal supporting grid was proposed by Trowell (1954). To prevent contact between the tissue and the metal, a piece of special paper or a layer of 2% agar is applied to the support (Trowell, 1959).

---

*Cultures of organs must not be confused with the methods of organ culture. In the latter case, the issue is not the character of the explanted material, but the properties of the culture method that provide the best facilities for specific differentiation.

(2) Culture on a lens paper raft (Chen, 1954). To prevent the raft from sinking, Richter (1958) suggested that the lens paper be coated with silicone.

(3) Culture on millipore filters. This method of organ culture, suggested and developed by Grobstein (1953, 1956), has several advantages; millipore filters with strictly defined pore size and thickness are used as the supporting substrate. A plastic ring, with the filter glued to it, is placed over the well containing the nutrient medium. One or two explants can be cultured on one filter.

(4) Culture on millipore filters by the multiple organ culture method enables many explants to be grown simultaneously on the same nutrient medium (Luriya and P'yanchenko, 1967). A Conway dish is used as the culture vessel; squares of millipore filters 49 mm² in area are held above the nutrient medium by a plastic platform with holes in it (Figs. 3 and 4).

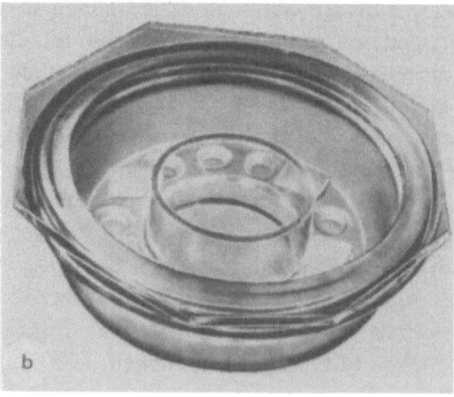

Fig. 3. Culture vessel for the multiple organ culture method: (a) disassembled; (b) assembled.

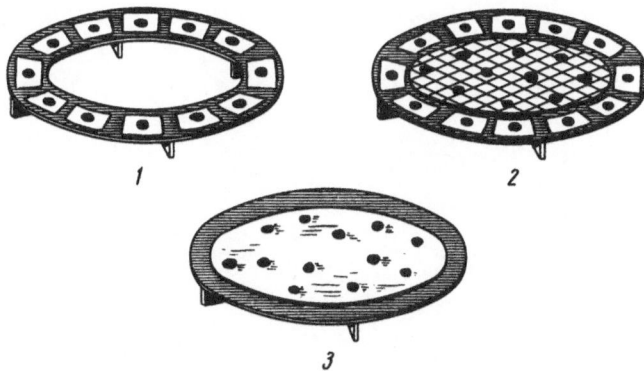

Fig. 4. The multiple organ culture method: (1) culture in Conway dish on small filters; (2) culture on large filter and on small filters in the same vessel (which allows filters of different types to be used and humoral tissue interactions to be studied); (3) culture on a large filter (allowing many cells to be obtained).

Practically all the methods of culture described above have been used for the explantation of lymphoid and hematopoietic tissue. The results have shown that tissues of such complex and heterogeneous composition behave differently depending on the technique of culture. By using these different methods of explantation, it is thus possible to observe lymphoid and hematopoietic tissues from various aspects. It is remarkable that the main differences among cultures of various lymphoid and hematopoietic tissues are determined by the technique of explantation used, rather than by the source of the actual tissues explanted. In culture, the common structural features shared by various hematopoietic and lymphoid tissues are often exhibited.

In this monograph, I shall analyze only the results obtained by culture of normal lymphoid and hematopoietic tissue under long-term conditions.

Much of the material in this book was obtained by the work of the author herself and of her colleagues in the Laboratory of Immunomorphology of the N. F. Gamaleya Institute of Epidemiology and Microbiology, Academy of Medical Sciences of the USSR. The author is grateful to A. Ya. Fridenshtein, Head of the Laboratory, for his critical remarks and his advice during the preparation of the manuscript.

CHAPTER I

Hematopoietic Tissue *In Vitro*

In the embryogenesis of higher vertebrates, hematopoietic cells initially appear in the yolk sac; later, foci of hematopoietic cells are also formed in the mesenchyme of the embryo. The next important step in the development of hematopoiesis occurs in the embryonic liver, after which it is transferred to the bone marrow, which throughout the rest of life serves as the principal hematopoietic organ (Metcalf and Moore, 1971). Each of these stages of development of hematopoietic tissue follows a different course if different explantation techniques are used.

1. Hematopoietic Tissue in Plasma Clot Culture

When the blastoderms of chick embryos at the 3–18 somite stage were cultured in a plasma clot, blood vessels developed from the mesenchymal cells in the form of isolated islets that later fused into a single system (McWhorter and Whipple, 1912).

Conversion of mesenchymal cells into hematopoietic cells, as well as the formation of blood vessels and islets of hematopoietic elements, could also be observed during culture of embryonic blastoderms in Locke-Lewis solution (Sabin, 1921). Although this method permitted intravital investigations of the cultures under high magnification, an important disadvantage was the low viability of the cultures. Observations in an individual culture were limited to 4–5 hr.

In cultures of the yolk sac of 6–8.5 day rabbit embryos, the process of formation of the hematopoietic elements was found to be the same as that in the yolk sac *in vivo* (Maksimov, 1909, 1925).

In cultures of human embryonic liver (from 1 to 3.5 months of intrauterine life), both granulopoiesis and erythropoiesis were observed (Benevolenskaya,

1929). In this instance, cultures were conducted in a plasma clot with the addition of an extract prepared from the embryo the liver of which was used for explantation. In some cultures, capillaries containing many erythrocytes were observed to develop. During the 1st week of culture, the successive stages of erythropoiesis could be followed. In the 2nd week of culture, the vessels contained only mature erythrocytes.

In cultures of the liver of a 12-day rabbit embryo in a plasma clot, Maksimov (1925) was able to maintain hematopoiesis for only a short time: on the 6th day *in vitro,* the liver parenchyma continued to proliferate and the bile ducts to grow, but erythroid cells were left only as small foci with signs of degenerative changes in their nuclei.

Plasma clot cultures enable hematopoiesis to be mantained in embryonic liver explants for only a short time. This period depends on the species of donor, the stage of embryogenesis at which the material is taken for culture, and also, possibly, on the composition of the nutrient medium (in particular, on the addition of homologous embryonic extract).

Several descriptions have been given of the culture of bone marrow in a plasma clot (Carrel and Burrows, 1910a; Foot, 1912; Maksimov, 1916; Erdmann, 1917; Rasmussen, 1933; Terent'eva, 1955).

Cultures of this type were first set up in 1910 by Carrel and Burrows; these were explants of the bone marrow of a kitten in cat plasma. After 24 hr, the fragments were surrounded by migrating erythrocytes and leukocytes; these cells later degenerated. Spindle-shaped cells then appeared in the zone of growth.

All investigators emphasize that hematopoietic cells in plasma clot cultures of bone marrow have a short life of only a few days. They are subsequently replaced by macrophages and proliferating fibroblast-like cells.

Do immature hematopoietic forms continue to differentiate into more mature cells (even for a short time) in such cultures, or can the bone marrow cells survive only for a short period *in vitro*? Conflicting answers have been obtained to this question. According to the observations of Maksimov (1916), progressive development and specialized differentiation of bone marrow cells do not take place in culture. Myelocytes remain viable for up to 5 days, after which some degenerate, while others become flatter and exhibit the property of phagocytosis. In Maksimov's opinion, the presence of transitional forms, i.e., of cells with the nucleus of a myelocyte and an extensive cytoplasm with phago-cytosed inclusions, points to the possibility of conversion of myelocytes into macrophages. Destruction of erythroblasts takes place earlier still.

On the other hand, according to Rasmussen (1933), myeloblasts can multi-ply in plasma clot cultures of bone marrow from young rabbits and may sometimes be converted into myelocytes. There is evidence that maturation of normoblasts takes place in culture, accompanied by expulsion of the nucleus (Van Herwerden, 1923; Spadafina, 1935).

The early stages of growth of rabbit bone marrow explants in plasma clot cultures were studied by Rachmilewitz and Rosin (1944). They set out to determine whether proliferation and differentiation of hematopoietic cells take place in such cultures. The ratio of immature to mature bone marrow cells 24 hr after explantation was shifted toward the mature cells in comparison with the original values. In the explant, the zone of mature erythrocytes was separate from the zone of myeloid hematopoiesis. After a few days, hematopoiesis disappeared and the explant became connective-tissue in character.

The changes that take place in the composition of hematopoietic tissue in bone marrow explants during culture can be explained in various ways. Plasma clot cultures are not a convenient model with which to study the conversion of some cell forms into others. For that reason, many of the observations on cell transformation in such cultures must be accepted with caution. On the basis of morphological observations, for instance, the possible sources of giant cells in bone marrow cultures could be fat cells, histiocytes, myelocytes, and other bone marrow cells.

Foot (1912) described the conversion of a large proportion of the small lymphocytes migrating from explants in chicken bone marrow cultures into large lymphocytes and later into myelocytes and polymorphs. However, these results could not be confirmed (Erdman, 1917). The same also applies to other observations of cell transformations in plasma cultures; various workers using similar tissues for study frequently failed to obtain mutually acceptable conclusions. Despite this indeterminacy of the results concerning cell conversions, the culture of bone marrow in plasma clots has yielded much valuable information. During cultivation of bone marrow in plasma clots, the hematopoietic cells persist in the first stage and for several days are capable of dividing and differentiating; later, they degenerate. The second stage of growth of the cultures is characterized by the appearance of macrophage–polyblasts. This is followed by proliferation of fibroblasts, which form a dense network around the explant. Whereas Maksimov's observations (1927) on the conversion of lymphocytes of the bone marrow into polyblasts were later confirmed in other models, and can be regarded as sufficiently conclusive, the question of the origin of fibroblasts in bone marrow cultures remains unanswered. Different methods of culture and special experimental analysis are required for the solution of this problem.

The data obtained by Reisner (1959) in submerged cultures of bone marrow fragments are in a class apart from the results of other investigations. Although these cultures were not plasma clot cultures, it is relevant to examine the results obtained by this method at this point.

Reisner states that hematopoiesis can be maintained for a long time (over a month) in cultures of human bone marrow fragments (Reisner, 1959). He found that growth of fibroblasts is prevented if the bone marrow is cultured under a deep layer of nutrient medium.

Fragments of bone marrow are explanted in a special vessel consisting of a cover slip and glass cylinder 18 mm in diameter affixed to it. The fragment of bone marrow is placed on the cover slip, which forms the base of the vessel, and is covered with a tantalum grid, which presses it against the glass. Nutrient medium containing plasma or serum is poured on from above. The depth of the layer of medium is considered to be the decisive factor in maintaining differentiation of the hematopoietic cells. For instance, if the layer of medium is 4 mm thick or less, growth of fibroblasts, macrophages, and giant cells is observed, whereas in "deep" cultures, growth of fibroblasts is suppressed, and the conditions favor the maintenance of hematopoiesis. In that case, as Reisner points out, differentiated bone marrow and blood cells migrate from the explants for several months. During the 1st month, all the elements of normal bone marrow including megakaryocytes migrate from it, while during the 2nd month, mature leukocytes are seen.

However, Farnes and Trobaugh (1961a,b), who used Reisner's technique in their experiments, did not confirm these observations. Fibroblast-like cells appeared regularly on the 2nd or 3rd day of culture, and after 1 week, they were the predominant cell type in the cultures. After 14 days, solitary hematopoietic cells appeared occasionally among the fibroblast-like cells. When collagenase was added to the medium, no fibroblast-like cells appeared, but the majority of cells on the 10th day of cultivation consisted of phagocytic macrophages and histiocytes. No appreciable prolongation of hematopoiesis was observed in the cultures under these circumstances.

Billen (1959) used Reisner's method to cultivate mouse bone marrow fragments. According to his observations, there is a gradual decrease in the number of immature hematopoietic cells and an increase in the number of macrophages and fibroblasts in the cultures. The decrease in the hematopoietic potential of the explanted cells correlates with these changes in the cell composition of the bone marrow cultures.

In like with these findings, injection of bone marrow from cultures into lethally irradiated animals gave beneficial results (preventing death from acute radiation sickness) until the 4th day of culture. The protective effect then diminished rapidly, and injection of cells cultured for 9 days was ineffective (Billen, 1959).

Reisner's method thus seems to have little future as a means of obtaining long-life cultures of hematopoietic tissue.

2. Monolayer Cultures of Hematopoietic Tissue

In monolayer cultures, not only can changes in cell morphology be studied, but also changes in the composition of the explanted cell population can be subjected to quantitative analysis. If bone marrow cells are grown in monolayer cultures, not all the transplanted cells become attached to the surface of the

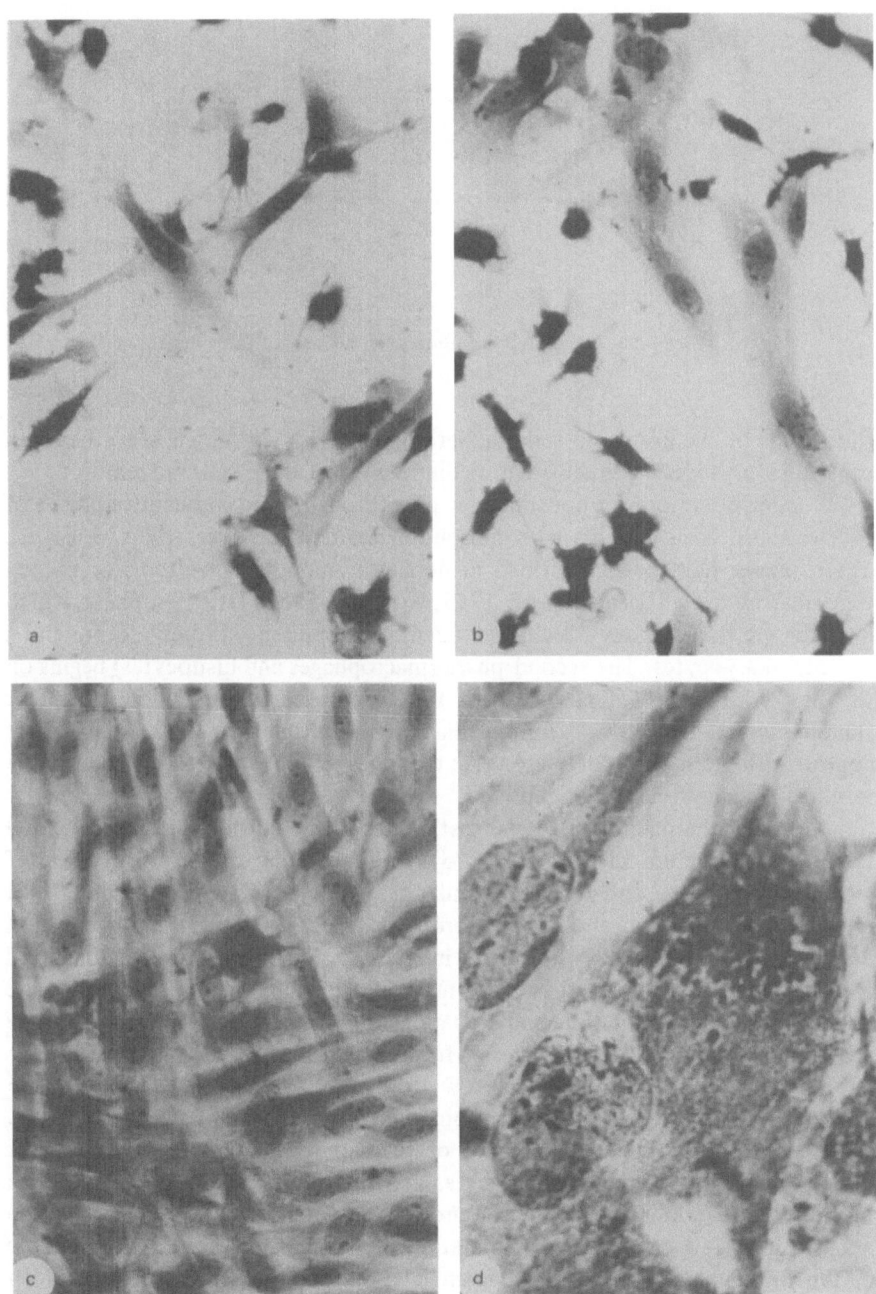

Fig. 5. Cell composition of monolayer cultures of guinea pig bone marrow: (a) 3 days; (b) 7 days; (c) 12 days; (d) 12 days. After Fridenshtein *et al.* (1970*b*).

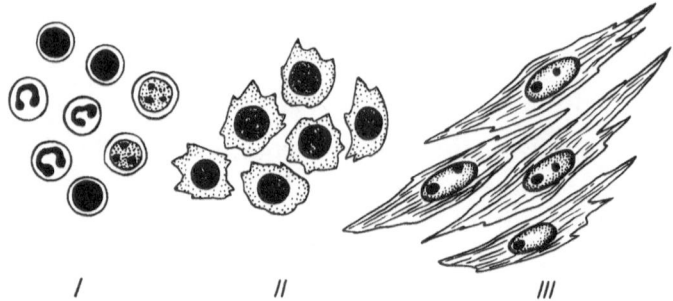

/   //   ///

Fig. 6. Three phases of growth in monolayer cultures of bone marrow cells.

slide. Cells of the erythroid series adhere only weakly to the glass, so that with subsequent changes of the nutrient medium, most of these cells are removed.

Monolayer cultures of bone marrow pass through three consecutive phases of development, each with its own morphological characteristics (Figs. 5 and 6). These phases have been described in detail for monolayer cultures of human bone marrow cells (Berman *et al.*, 1955; Woodliff, 1958). The first phase, which lasts for the first few days, is characterized by the presence of polymorphs and a few megakaryocytes. The second phase (macrophages and histiocytes) begins on the 2nd or 3rd day of culture with the appearance of round, polygonal, or spindle-shaped cells 20–40 μm in diameter. Multiple vacuoles and granules appear in the cytoplasm of these cells; they are probably material formed as the result of phagocytosis of cell debris.

As a rule, the macrophages and histiocytes in the cultures are not collected into groups, but are dispersed (Fig. 5). The beginning of the third phase of growth overlaps the end of the second phase. Cells characteristic of the third phase and with the morphology of fibroblasts appear between the 6th and 20th days (Figs. 5 and 6). They proliferate intensively and form extensive zones and foci. In their morphology, these cells resemble cells of diploid strains; they are easily removed from the slide and contain tonofibrils in the cytoplasm. All these characteristics distinguish them clearly from histiocytes.

During cultivation of bone marrow from different species of animals, the same phases of growth are observed, but their duration may vary considerably. For instance, after the disappearance of the hematopoietic cells in cultures of mouse and rat bone marrow, histiocytes and fibroblasts, i.e., cells characteristic of the second and third phases of growth, respectively, are present simultaneously for a few days. A pure population of fibroblast-like cells, characteristic of the third phase, appears only much later—after several months (McCulloch and Parker, 1957).

On the other hand, if guinea pigs are used as the donors, a population of intensively proliferating fibroblasts is found 10–12 days after explantation. Foci of fibroblast-like cells (Fig. 7) can be seen at this time against the background of the histiocyte population detaching itself from the slide (Chailakhyan, 1970). Hematopoietic cells thus remain attached to the slide only for a short time in monolayer cultures of bone marrow. What is their subsequent fate? Most of them degenerate; some of the immature hematopoietic cells may manage to differentiate under these circumstances into mature forms. But usually,

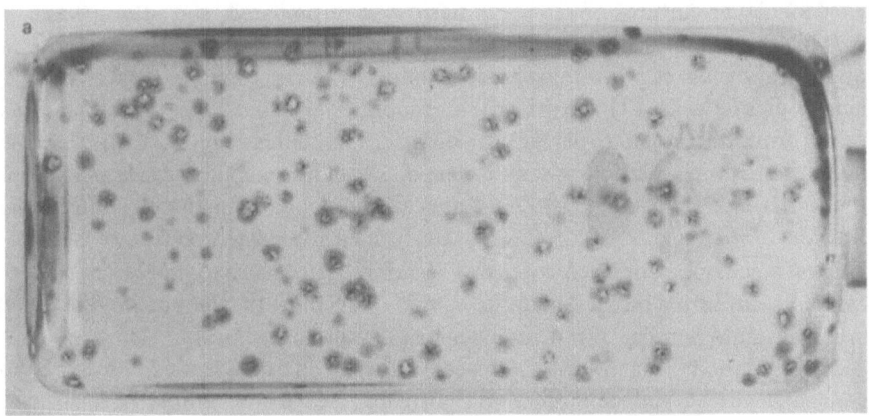

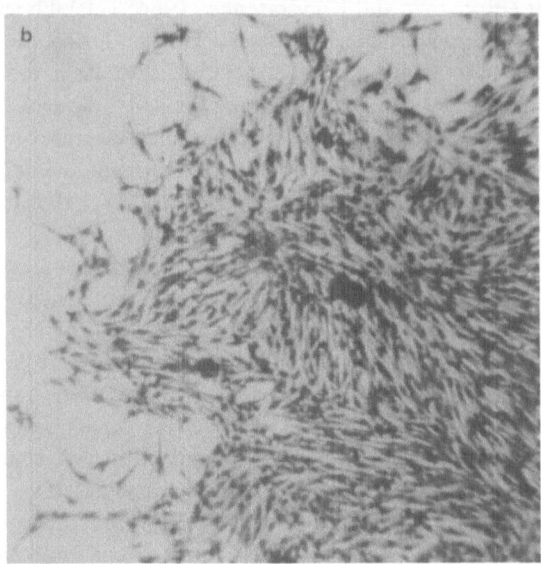

Fig. 7. Colonies of fibroblasts in 15-day bone marrow culture. After Fridenshtein *et al.* (1970*b*).

monolayer cultures do not enable hematopoiesis to be maintained for a long time *in vitro*.

The sources of the cells populating the bone marrow culture in the second phase are another problem. Do these cells arise entirely through division of the macrophage–histiocytes present in the original explant, or are hematopoietic cells also involved in their formation?

An attempt to answer this question was made by studying the kinetics of appearance of the cells in the second phase in monolayer cultures of rat bone marrow (Luriya *et al.*, 1966*a*).

Monolayer cultures of rat bone marrow were prepared, and after 70 min to 120 hr, the number of nucleated cells was counted and the relative percentages of the following cell forms were determined: (1) macrophage–histiocytes; (2) hematopoietic cells; (3) mature polymorphonuclear leukocytes; (4) degenerated cells. From the relative percentage of macrophage–histiocytes, their absolute number in the population was determined. It can be concluded from analysis of the kinetics of the change in the number of macrophage–histiocytes during the first 24 hr that many of these cells are formed by transformation of hemato-poietic precursors, for even intensive proliferation of macrophages and histio-cytes present in the original material could not account for the observed increase in the absolute number of macrophage–histiocytes in the culture.

Investigations claim that the cells arising in the second phase from connective tissue and hematopoietic elements may differ morphologically and histochemically (Biryuzova and Kondratenko, 1964; Terskikh and Kondratenko, 1962; Rappay *et al.*, 1968; Fazekas *et al.*, 1968). During the first hours and days, many cells with stratified and branching cytoplasm but with a ring- or horseshoe-shaped nucleus can be seen among the cells adherent to the glass in rat bone marrow cultures. The cells fixed to the glass are also heterogeneous on histochemical analysis. For instance, about one-third of the cells give a positive reaction for peroxidase and alkaline phosphatase; i.e., they contain enzymes characteristic of myeloid cells (Rappay *et al.*, 1968).

Virolainen and Defendi (1968) found that the macrophage–histiocytes in cultures are descendants of hematopoietic stem cells. These results were obtained by culturing hematopoietic foci from the spleens of 13–14 day radiation chimeras, restored by transplantation of bone marrow from donors with a marker chromosome ($T_6 T_6$). Hematopoietic foci from the radiation chimeras were grown on grids in organ cultures. A slide was placed on the bottom of the vessel, and cells migrating from the explant fell on it. After culture for 3 or 4 days, these cells were subjected to chromosome analysis. This analysis showed that all dividing cells contained the marker ($T_6 T_6$), which was evidence of their origin from the cells of the donor's bone marrow. It is important to note that cells giving rise to macrophages migrate from granulocytic and mixed splenic colonies, whereas erythroid colonies do not contain the precursors for these forms. Since each colony is a cell clone arising from a hematopoietic stem cell, it can of course be concluded that precursor cells for macrophages present in the granulocytic colonies are also descendants of hematopoietic stem cells.

To study the dynamics of formation of the fibroblast-like cells characteristic of the third phase of development of the bone marrow explants, cultures from guinea pigs, rabbits, and man provide excellent models. In monolayer cultures of guinea pig bone marrow, most of the cells after 24 hr are represented by macrophages and polymorphs. After 3 days, single large, elongated cells or cells split lengthwise into two, with a large oval nucleus containing several nucleoli, begin to appear among cells of macrophage–histiocyte type. By the 6th day, separate foci of 4–8 fibroblast-like cells can be seen. At this time, the leukocytes have almost completely disappeared. Most of the cells in the culture, as before, consist of small histiocytes with short, branching processes. Later, the number of fibroblast-like cells increases considerably, and on the 9th day, compact and scattered foci consisting of 50–80 cells are formed. On the 12th day, large foci of fibroblasts with tonofibrils clearly distinguishable in their cytoplasm can be seen. Later, these foci join together. By the 20th day, fibroblasts are the predominant cells in the cultures (Fridenshtein *et al.*, 1970*a–c*).

Fibroblast counts showed that from the 3rd to the 6th day, the number of cells quadruples, after which it rises slowly until the 20th day (Chailakhyan, 1970). This finding suggests, of course, that the fibroblast population increases by proliferation of the fibroblasts present in small numbers in the 3-day cultures, and that between the 3rd and 6th days, these cells divide intensively, with a doubling time on the order of 20 hr (Fridenshtein *et al.*, 1970*b*). Fibroblast-like cells in cultures of hematopoietic tissue possess high proliferative activity. That they do was first pointed out by Maksimov (1928). Fibroblast colonies in monolayer cultures of guinea pig bone marrow and spleen also contain many dividing cells, and one can expect to find a high proliferative pool in them.

The study of incorporation of H^3-thymidine into fibroblasts in bone marrow cultures supports this hypothesis. The index of pulse labeling in 7-day cultures is 18%, and saturation arises after 32 hr, when the labeling index reaches 90%. In 12-day cultures, with the same pulse-labeling index, saturation takes place between 48 and 72 hr, when the labeling index is 78% (Keilis-Borok *et al.*, 1971). It thus follows that the proliferating pool is extremely high in colonies of fibroblasts in 7-day cultures, and it decreases slightly by the 12th day of cultivation. This decrease may occur because some of the cells in these colonies are undergoing differentiation, and because of this differentiation they cease to divide. The mean time of generation of fibroblasts in colonies, determined from the saturation curves, is about 20 hr for 7- and 12-day cultures.

Cells composing the colonies are typical fibroblasts, near which collagen fibers are formed. In their morphology, these cells are identical to fibroblasts arising in peripheral blood cultures in a plasma clot and studied in detail by Maksimov. Histochemical investigation showed that these cells react positively for acid phosphatase and negatively for alkaline phosphatase. Fibroblasts are characteristic of the third phase of growth of monolayer bone marrow cultures from guinea pigs, rabbits, man, rats, and mice. These cells are detached from the slide with trypsin, and can be subcultured repeatedly. Under these conditions, they do not lose their typical morphology, and they retain a diploid set of chromosomes (Kokorin *et al.,* 1970; Panasyuk *et al.,* 1972).

In order for discrete colonies of fibroblasts to form after primary explantation of bone marrow cells, the number of explanted cells had to be not less than 10^5 and not more than $2-3 \times 10^5$ per square centimeter of the base of the vessel. Most cells in culture (leukocytes, macrophages) under these conditions do not themselves take part in colony formation, but play the role of natural feeder cells for the colony-forming cells. If the density of initial explantation is lower, bone marrow cells irradiated at a dose of 4000 R can be used as an additional source of feeder cells. Each colony of fibroblasts arises from a single fibroblast colony-forming cell (FCFC); i.e., the colony is a clone (see below). For this reason, the number of colonies reflects the number of precursor cells for fibroblasts among the cells explanted. This number is about 10^{-5} for guinea pig and human bone marrow (Fridenshtein *et al.,* 1970*a–c*; Luriya *et al.,* 1972*a*). This natural type of cloning takes place because the population contains many cells that act as natural feeder cells for fibroblast development during culture.

Chromosome analysis of colonies in mixed cultures of marrow cells from male and female guinea pigs showed that all metaphases in each colony belong entirely to the male karyotype or to the female karyotype; no mixed foci were

TABLE 1. Results of Chromosome Analysis of Fibroblast Foci in Mixed Cultures of Bone Marrow Cells of Male and Female Guinea Pigs[a]

Expt. No.	Number of identified metaphases in colonies	Male/Female
1	4	0/4
2	4	0/4
3	5	5/0
4	3	3/0
5	3	0/3
6	4	4/0
7	4	4/0
8	3	0/3
9	5	0/5
10	4	4/0

[a]After Chailakhyan *et al.* (1970).

found (Table 1). It can accordingly be concluded that each colony of fibroblast-like cells is a clone that develops from a single FCFC (Fridenshtein *et al.*, 1970*a–c*).

This conclusion is in good agreement with the linear relationship found between the number of explanted cells and the number of colonies of fibroblasts formed in the culture. This relationship holds true if the number of explanted cells is between 10^6 and 10^7 per 100 ml flask (Fig. 8). The size of the colony-forming unit, i.e., the number of explanted cells for each colony, is about 10^5 nucleated cells for guinea pig and human bone marrow.

Probably not all colony-forming cells, i.e., precursors of fibroblasts, give rise to colonies under culture conditions, and their actual concentration in the bone marrow suspensions is therefore perhaps higher than the concentration detected by colony formation *in vitro*.

Further evidence of the clonal nature of fibroblast colonies in bone marrow cultures was obtained by thymidine-H^3 labeling (Keilis-Borok *et al.*, 1971). If thymidine-H^3 is present during only part of the time needed for all the FCFCs to pass through the first period of DNA synthesis *in vitro* (which corresponds to

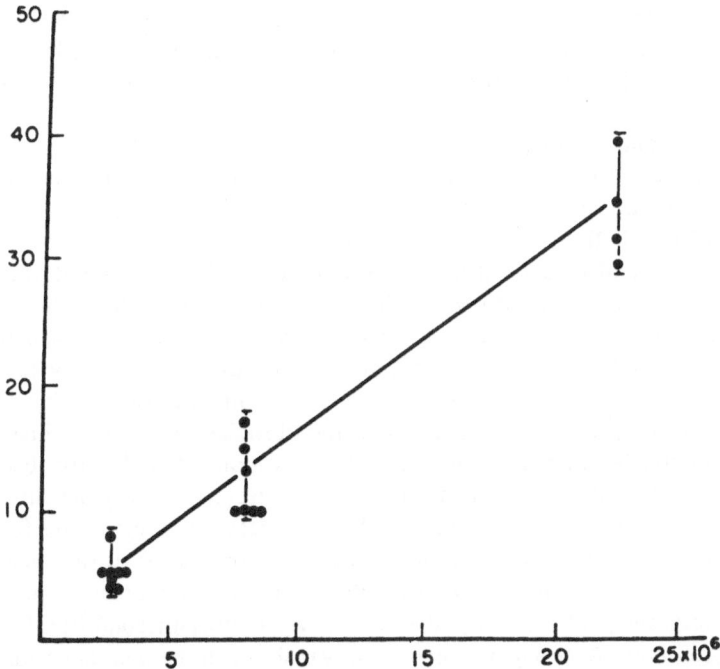

Fig. 8. Number of colonies of fibroblasts as a function of number of explanted bone marrow cells. Abscissa: number of explanted cells; ordinate: number of colonies growing. After Fridenshtein *et al.* (1970*b*).

28–60 hr), and cultivation continues later with the addition of nonlabeled thymidine, some of the colony-forming cells will be labeled, while others will not. If each colony is a cell clone, some colonies must thus be entirely labeled and others entirely unlabeled. Mixed colonies consisting of labeled and unlabeled cells can only be exceptions that depend on the random amalgamation of several colony-forming cells. The results that were obtained provide definite evidence in support of the clonal nature of fibroblast colonies: of 97 colonies, 94 consisted entirely of either labeled or unlabeled cells.

Precursor cells for the colonies of fibroblasts arising in cultures are outside the proliferating pool in the original bone marrow.

If thymidine-H^3 was injected into guinea pigs and the bone marrow was subsequently explanted to form monolayer cultures in a medium containing unlabeled thymidine, the overwhelming majority of fibroblasts were found to be unlabeled. Even if thymidine-H^3 was administered in saturation *in vivo* (every 2 hr for 3 days), the index of labeled fibroblasts in the culture was only 2–5%, whereas the index of labeled histiocytes exceeded 50%.

In bone marrow cultures from an animal aged 6 days, about 15% of fibroblasts were labeled after injection of saturating doses of thymidine-H^3 *in vivo*. Judging from the low intensity of incorporation of thymidine-H^3, precursors of fibroblasts thus belong to the category of bone marrow cells that are at rest in a state of equilibrium or have a very long life cycle (Keilis-Borok *et al.*, 1972).

Cells forming colonies of fibroblasts (FCFCs) gradually enter into the S-period 28 hr after explantation. In fact, after addition of thymidine-H^3 to bone marrow cultures for the first 4–72 hr and subsequent growth on medium with nonlabeled thymidine, the labeling index of fibroblasts in 3–5 day cultures was found to depend on the duration of the initial period during which the cultures were grown with thymidine-H^3 in the medium (Keilis-Borok *et al.*, 1971). If thymidine-H^3 was present in the medium for not more than the first 28 hr, the fibroblasts were unlabeled, although many of the histiocytes in these cultures contained the label. If thymidine-H^3 was present in the medium for the first 31 hr or longer, the labeling index of the fibroblasts was 30% or higher. The increase in the labeling index in this case corresponded to lengthening of the time of contact of the cultures with the thymidine-H^3. The labeling index reached a maximum (98%) if thymidine-H^3 was present in the cultures for 60 hr. These results show that incorporation of label into precursors of fibroblasts does not take place during the first 28 hr of culture, and that the precursor cells enter the proliferating pool gradually after 28 hr. An important condition for initiation of proliferation among precursor cells was shown to be their adhesion to the surface of the substrate. When bone marrow cells suspended in the medium were incubated in the presence of thymidine-H^3, even after 24 hr, the fibroblasts contained virtually no label (Keilis-Borok *et al.*, 1971). This property constitutes

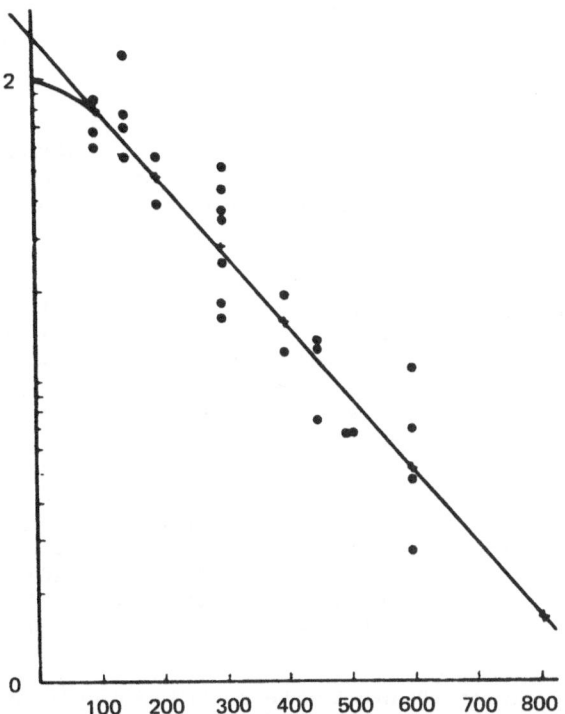

Fig. 9. Curve showing survival of bone marrow cells forming colonies of fibroblasts, after irradiation *in vitro*. Abscissa: dose of irradiation (in R); ordinate: log of percentage of surviving FCFCs. D_0 178 R; $r = 1.44$. After Kuz'menko *et al.* (1972).

an essential distinction between precursors of fibroblasts and precursors of histiocytes. The latter begin to proliferate from the time of explantation, and proliferative activity in bone marrow explants reaches its maximum in the first 24 hr of cultivation.

Precursors of fibroblasts were found to be highly radiosensitive. When an original suspension of human bone marrow cells was irradiated *in vitro* at a dose of 600 R, followed by explantation into monolayer cultures, fibroblast colony formation was sharply inhibited (Panasyuk *et al.*, 1972).

The survival curve of FCFCs as a function of the dose of irradiation (Kuz'menko *et al.*, 1972) of guinea pig bone marrow cells *in vitro* is shown in Fig. 9. This curve shows that D_0 for FCFC is 178±15 R.

The FCFCs are highly adhesive cells (Fridenshtein, 1973). They virtually all adhere to the surface of the glass in 60–90 min (Table 2 and Fig. 10).

There is experimental evidence of sharp changes in the concentration of FCFCs in bone marrow as a result of procedures that upset the state of

TABLE 2. Adhesiveness of Bone Marrow and Thymus Cells Forming Colonies of Fibroblasts in Monolayer Cultures[a]

Expt. No.	Explanted material	Time from explantation to removal of nonadherent cells[b]	Number of explanted cells	Number of colonies in culture on 12th day	Cloning efficiency (%)[c]
1	Guinea pig bone marrow	0–48 hr[d]	10^7	462, 403	100
		0–90 min[d]	10^7	424, 356	90
		0–30 min[d]	10^7	435, 387	95
		30–60 min[e]	Not determined	56, 31	10
		60–90 min[f]	Not determined	4, 4	1
2	Guinea pig thymus	0–48 hr[d]	5×10^7	40, 53	100
		0–90 min[d]	5×10^7	43, 47	97
		0–30 min[d]	5×10^7	44, 53	104
		30–60 min[e]	Not determined	0, 5	5
		60–90 min[f]	Not determined	1, 4	5
3	Rabbit thymus	0–48 hr[d]	2×10^7	117, 173	100
		0–30 min[d]	2×10^7	100, 131	79
		30–60 min[e]	Not determined	9, 14	8
		60–90 min[f]	Not determined	0, 0	0

[a]After Latsinik and Epikhina.
[b]Bone marrow or thymus cells were introduced into 250-ml Roux flasks. The cells were given different times to adhere, after which the medium was changed and 10^7 cells of the irradiated feeder were added to each flask.
[c]Relative to culture with cells allowed to adhere for 48 hr.
[d]Time of adhesion of primary explanted cells.
[e]Primary explanted cells not adherent after the first 30 min transferred to a new flask.
[f]Primary explanted cells not adherent after the first 60 min transferred to a new flask.

equilibrium of the hematopoietic tissue, and in particular, after sublethal X-ray irradiation and blood loss. If guinea pig or mouse bone marrow is explanted immediately after irradiation at a dose of 150 R, the number of FCFCs will be reduced by not more than 10%. The same result is found for the spleens of these animals. However, 5 and 10 days after irradiation, the number of FCFCs in the mouse and guinea pig bone marrow is reduced by 10 times; conversely, in the spleens of irradiated mice on the 5th–20th day after irradiation at a dose of 150 R, the number of FCFCs is increased by 5–10 times (Kuz'menko *et al.*, 1972). The number of FCFCs in guinea pig bone marrow 2 hr after blood loss is increased several time (Fridenshtein *et al.*, 1974*b*).

The number of bone marrow FCFCs detected by the *in vitro* colony-assay method, being relatively constant in normal animals, thus changes sharply under the influence of factors disturbing the steady state of hematopoietic tissue. This finding may indicate an essential role of these cells in hematopoiesis.

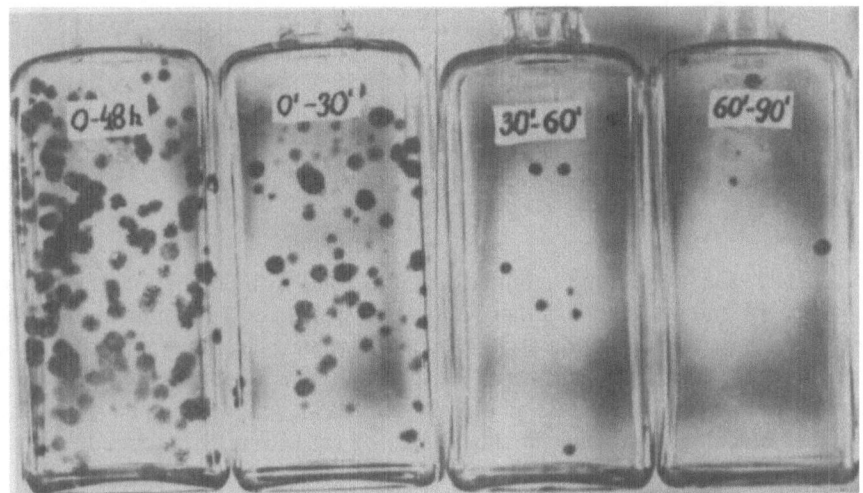

Fig. 10. Fibroblast colony formation in cultures of bone marrow cells subjected to various conditions of adhesion. Explanation in Table 2. After Fridenshtein (1973).

By the 2nd day, individual fibroblasts can be seen in monolayer cultures of embryonic mouse liver, a hematopoietic organ in the embryo, and in 4-day cultures, they form colonies. By the 14th day, these colonies attain a large size and are easily distinguished by the unaided eye. Depending on the closeness of packing of their cells, the colonies can be divided into more compact ones and loose ones. They consist of fibroblasts with typical morphology. The colony-forming unit consists of about 10^5 nucleated cells. Unlike histiocytes in bone marrow cultures, histiocytes in embryonic liver cultures persist for a long time, meanwhile exhibiting high proliferative activity (Latsinik and Keilis-Borok, 1971).

Fibroblasts from bone marrow cultures are easily removed from the slide by the action of trypsin and can be subcultured. By repeated subculture, such fibroblasts can easily be obtained in large numbers (Fig. 11).

In fibroblast colonies growing in the bone marrow cultures, characteristic changes take place in the proportion of fibroblasts and FCFCs if embryonal serum is not added to the medium. The ratio between these two categories of cells in successive stages of cultivation of guinea pig marrow is shown in Fig. 12. It is clear that before the 11th day of cultivation, the total number of cells in culture rises exponentially and doubles in about 19–22 hr, while between 11 and 18 days, the increase in the total fibroblast population in the culture slows down, and the time it takes to double is 42–50 hr. Meanwhile, the increase in the FCFC population in the cultures occurs at a constant rate between 3 and 18 days. The time required for the number of these cells to double is 42 hr, i.e.,

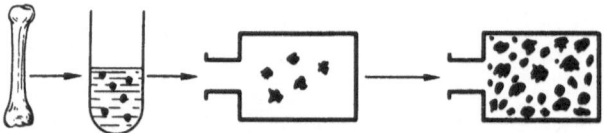

Fig. 11. Scheme showing increased efficiency of colony formation during subculture of cells from monlayer bone marrow cultures.

practically the same time required for the total cell population to double between 11 and 18 days. In other words, between 3 and 7 days, the number of FCFCs as a percentage of the total cell population in the colonies gradually falls, and on the 11th day, it reaches a level of 2%, where it remains until the 18th day. The number of FCFCs in this investigation was determined from the number of colonies growing in subcultures of cells harvested from cultures at

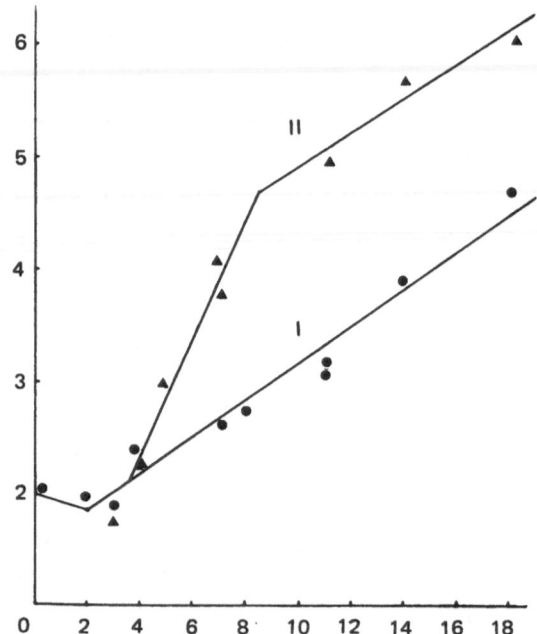

Fig. 12. Change in number of colony-forming cells (I) and in total number of fibroblasts (II) in cultures at different times after explantation. Abscissa: age of cultures (days); ordinate: log of percentage of number of fibroblasts and of number of colony-forming cells. After Fridenshtein (1973).

different times after explantation. The number of colonies formed prior to harvesting gave an indication of the number of FCFCs in the initially explanted cell population. Subculturing cells taken on the 2nd and 3rd day after primary explantation led to the formation of virtually the same number of fibroblast colonies as was observed in the primary (not subcultured) explants. It thus follows that the efficiency of subculture of FCFCs in this particular system was very high. The change in the content of FCFCs during subculture of cells from cultures obtained at different times thus reflects a change in the concentration of colony-forming cells in the course of development of the colonies. It follows from Fig. 12 that during the first 3 days, as many FCFCs remain in the cultures as were introduced during primary bone marrow cell explantation. The number of FCFCs then begins to increase, and a population of fibroblasts without FCFC properties appears among the cells of the colonies. The growth in the fibroblast population between the 3rd and 11th days takes place significantly faster than the growth in the FCFC population. Starting from the 11th day, however, a state of equilibrium arises: although the colonies continue to increase in size, their FCFC content is maintained at a constant level on the order of 2–4%. Even so, only a few cells in the colonies are capable of self-support, i.e., of forming new colonies on subculture. On the other hand, more than 80% of cells in 7- and 12-day colonies are in the proliferating pool (see p. 155).

These results suggest that during the development of fibroblast colonies in the absence of embryonal serum, a line of colony-forming cells is maintained within clones. At the same time, these cologenic cells seem to serve as a source for repeated formation of populations of rapidly proliferating fibroblasts incapable of maintaining themselves for a long time.

That the curves for the FCFC population and the total population of fibroblasts are parallel and lie some distance apart on the graph can evidently be explained by assuming that the rapidly proliferating fibroblasts that are descended from the FCFC pass through about six divisions between the 3rd and 11th days in culture, as a result of which about 64 cells are formed from each one (this number corresponds to the number of T-fibroblasts determined in the colonies on the basis of thymidine-H^3 uptake). The concentration of FCFC among all the cells of the colonies is thus also about 2%.

3. Suspension Cultures of Hematopoietic Tissue

The essence of the method of suspension culture in its various modifications is that the cells are kept in a suspended state by continuous rotation, shaking, stirring, or bubbling of a gas mixture through the medium, so that they do not adhere to the surface of the culture vessel. The suspension culture method as applied to bone marrow and blood cells has not yet shown prospects for

prolonged maintenance of differentiation *in vitro*. The normal morphology of
the bone marrow cells is preserved for a short time (12–24 hr) in the cultures;
later, hematopoiesis and progressive maturation of the hematopoietic cells cease
(Lau *et al.*, 1967).

With the suspension culture method, it is possible to study the absolute
number of cells during growth by successive sampling. The method of bone
marrow culture most widely used at present employs vessels used in vaccine
production, into which a suspension of bone marrow cells in nutrient medium
containing 30–60% serum and glucose is poured.

Since the time for which hematopoiesis can be maintained in suspension
cultures is limited to a few days, this method can be used for short-term
experiments aimed at studying certain mechanisms of regulation of hematopoiesis and the metabolism of hematopoietic cells, as well as cytological and
autoradiographic analysis of hematopoietic tissue. Suspension cultures of bone
marrow have been used, in particular, to study hemaglobin synthesis on the basis
of radioactive iron incorporation, and the effect of erythropoietin on erythropoiesis (Shekhter, 1965).

Important results used for constructing models of proliferation and maturation of hematopoietic cells (including human) have been obtained by means of
the short-term suspension culture method combined with autoradiography. The
maturation time of myeloid and erythroid forms of bone marrow determined in
suspension cultures corresponds to the parameters for hematopoiesis in the
intact organism. The mean time for differentiation from myelocyte to polymorph in human bone marrow suspension cultures is 48–50 hr, while the
maturation time from erythroblast to normoblast is 48 hr (Kozinets, 1962).
Examination of these results, which are of great interest for the construction of
models of hematopoiesis, is beyond the scope of this book. The appropriate data
are described in several surveys, to which the reader is referred (Kozinets, 1962;
Lajtha, 1965; Bond *et al.*, 1965).

4. Formation of Hematopoietic Foci in Agar Cultures

In 1965, Pluznik and Sachs (1965) suggested a method of obtaining colonies
from suspensions of spleen cells by culture in agar on a feeder layer of mouse
embryonic cells. Two types of foci developed in the cultures—one consisting of
large cells with metachromatic granulation, the other containing neutrophilic
granulocytes. Cells with metachromatic granules were described by these
workers as mast cells. Electron-microscopic analysis subsequently showed, however, that these cells are macrophages containing phagocytosed particles of agar.
They have uneven surfaces and nuclei of different shapes. There are microvilli on
the cell surfaces, and the cytoplasm contains numerous inclusions and vacuoles
filled with amorphous material. The rough and smooth endoplasmic reticulum is

present in certain areas, mitochondria are numerous, and the Golgi apparatus is well developed.

Colonies of macrophages and neutrophils were formed from adult mouse spleen cells and embryonic mouse liver cells by cultivation in agar on a feeder layer of mouse embryonic cells or by the use of a conditioned medium (Pluznik and Sachs, 1966; Ichikawa *et al.*, 1967). The conditioned medium was produced by cultivating 10^6 mouse embryonic cells in monolayer cultures in 100-ml dishes for 7 days. The addition of 25% conditioned medium to agar cultures of spleen cells led to the formation of colonies similar to those that develop during growth on a feeder layer of embryonic cells (Pluznik and Sachs, 1966).

Other workers later showed that mouse bone marrow cells form colonies of myelocytes and macrophages when trypsinized kidney cells from 8-day mice are used as the feeder layer (Bradley and Metcalf, 1966). They found that only cultures of trypsinized mouse kidneys and embryonic cells could produce a suitable conditioned medium, whereas the medium from other cultures (lung, brain, or suspension of thymus, spleen, or liver cells) was ineffective (Bradley and Sumner, 1968).

The formation of colonies of myeloid cells and macrophages is thus observed during culture of a suspension of hematopoietic tissue on appropriate feeder layers or by the use of a conditioned medium (Metcalf and Moore, 1971).

The methods suggested by Pluznik and Sachs (1965) and Bradley and Metcalf (1966) have much in common. They are based on the use of a culture medium consisting of a double layer of agar. The bottom layer of 0.5% agar contains the trypsinized feeder cells; the top layer of 0.3% agar, the suspension of hematopoietic cells. The main components of the nutrient medium was Eagle's medium, to which various additional materials and serum were added. Cultivation is carried out in an atmosphere of 5–10% carbon dioxide in air. The methods used by these workers differed in certain technical details: the size of the plastic dishes for the cultures, the cell density, the sera used, and the methods of preparing and adding the conditioned medium.

When feeder layers consisting of living cells are used, the composition and development of the colonies show certain special features. For example, on a feeder layer of kidney cells, bone marrow cells form colonies that are visible on the 2nd day and reach their maximal size on the 7th–10th day of culture. These are the largest colonies. The greatest number of cells forming a colony is 2000–4000. These cells are mainly of the myeloid series, with a few macrophages. Later, macrophages begin to predominate, and purely macrophagal colonies are formed, although myeloid hematopoietic cells are found until the 14th day of cultivation (Bradley and Metcalf, 1966; Metcalf *et al.*, 1967).

A feeder layer derived from the kidneys of newborn mice and mice aged 2, 8, or 14 days possesses the ability to stimulate colony formation. This activity decreases with an increase in the donor's age. From 400 to 500 colonies are formed per 10^6 bone marrow cells on a feeder layer of kidney cells from 14-day mice, while if the donor's age is 2 months, the mean number of colonies is 340. Kidney cells from 3-month mice form on the average 45 colonies per million nucleated bone marrow cells. No colonies are formed in bone marrow cultures without the feeder layer.

In cultures of dissociated spleen cells on a feeder layer of trypsinized embryonic cells, the colonies reach their greatest size on the 10th day of growth; they consist on the average of 1200–1300 cells. The cell population in the colony remains at this level until the 17th day. Most colonies observed from the 3rd to the 10th day or later consist of actively phagocytic macrophages (Pluznik and Sachs, 1965).

Somewhat different results were obtained by the use of conditioned media, although in such cases also, colonies of myeloid cells and macrophages could be obtained (Pluznik and Sachs, 1966; Bradley, 1968a,b; Bradley and Sumner, 1968). Bradley and Sumner obtained conditioned medium from monolayer cultures of trypsinized kidneys of 8-day mice and by cultivating dissociated cells of mouse embryos aged 16–18 days. When conditioned medium from kidney cells was added, more colonies with myeloid differentiation were formed in agar cultures than by the use of the mouse embryo cell medium. Optimal conditions for the formation of conditioned medium were provided by culturing 20×10^6 cells in 5 ml medium for 6 days. The dose of conditioned medium added to agar cultures of bone marrow was 0.5 ml. Colonies of two types were seen to be formed in the cultures: (1) foci consisting of cells of the myeloid series, including metamyelocytes with ring- and horseshoe-shaped nuclei; and (2) colonies consisting mainly of monocytes. As cultivation continued, colonies of macrophagal type became predominant.

The activity of the conditioned medium is not reached by dialysis. The factor contained in the conditioned medium is resistant to heating; it withstands 50–80°C for 30 min (Bradley and Sumner, 1968).

It is interesting to note that conditioned medium can be produced by mouse embryonic cells irradiated at a dose of 4000 R (Pluznik and Sachs, 1966). These workers found that the factor (or factors) responsible for colony formation is thermostable and withstands prolonged heating (7 days at 37°C) without loss of activity, and is not inactivated by ribonuclease, although it is affected by dialysis. Dialysis of conditioned medium for 24 hr in the cold reduces its activity almost by half. Resistance to heat, consequently, has been confirmed by all investigators. Differences of opinion regarding the action of dialysis on the activity of the medium are probably attributable to differences in the techniques used.

Subsequent experiments showed that the ability to stimulate colony formation in agar cultures of bone marrow is possessed by factors other than conditioned media. This has been demonstrated most clearly by the addition of serum from mice with spontaneous or transplanted leukemia and serum from leukemic patients to cultures of bone marrow cells (Bradley *et al.,* 1967; Metcalf and Foster, 1967; Foster *et al.,* 1968; Stanley *et al.,* 1968).

Bradley *et al.* (1967) showed that blood serum from Akr mice with spontaneous lymphatic leukemia stimulates colony formation in suspensions of mouse bone marrow cells. Microcolonies appear in the stimulated cultures after only 2 days, and by the 6th day, large colonies containing up to 200 cells are found. Later, the rate of growth decreases. The colonies reach their largest size (600 cells) by the 10th day.

Cytological analysis of the colonies showed that on the 2nd day, they consist of 4–12 large cells with ring- or horseshoe-shaped nuclei, and also of smaller cells with similarly shaped nuclei. After 3 days, mononuclear forms of macrophages, phagocytosing agar, predominate in the colonies. On the 5th day, the myeloid cells are completely replaced by macrophages. At this time, round mononuclear cells with nuclei measuring 4–5 μm begin to appear.

The serum of mice with primary leukemia has a similar effect on cultures.

During the first 2 days, microcolonies were observed in the control cultures, but they did not subsequently grow in size, and most disappeared during cultivation. In some cases, the stimulating factor was found in the serum of healthy mice, but its activity was much lower (Metcalf and Foster, 1967). A study of the properties of this factor showed that it is thermolabile; that it is resistant to the action of ether, deoxyribonuclease, and ribonuclease; and that it consists of undialyzed molecules. On electrophoresis, it migrates with the α-globulin fraction (immediately after albumin). Stanley *et al.* (1968) consider that this substance is probably a protein or glycoprotein with specific leukopoietic activity.

A factor that stimulates colony formation is also present in the urine of healthy persons. On the 7th day of incubation, colonies in treated bone marrow cultures consist entirely of mononuclear cells.

Rat bone marrow does not yield colonies on a feeder layer of neonatal rat and mouse kidney. Colony formation was obtained on the addition of rat serum to the nutrient medium. Simultaneous addition of rat serum with other sera (calf or bovine) gave the most effective stimulation. Large colonies were produced in which myeloid cells persisted until the 6th day, but later, only monocytes were present (Bradley and Siemienowicz, 1968). These workers consider that monocytes are transformation products of myeloid cells.

The colony formation that takes place under the influence of the various stimulating factors contained in serum and urine thus differs from colony formation on a feeder layer. In the former, the length of life of the myeloid cells is shorter than in the latter, and smaller colonies are formed.

The colony-forming activity of human bone marrow cells can be studied by the agar culture method. However, the use of feeder layers and the addition of conditioned media to stimulate the formation of hematopoietic foci from rodent bone marrow and spleen do not provide conditions sufficiently favorable for colony development from human bone marrow. With the use of mouse kidney cells as a feeder layer, the number of cells in human bone marrow colonies did not exceed 50. It was later found that a conditioned medium obtained from cultures of human spleen cells stimulates colony formation. The addition of such a medium in a proportion of 25% to the bottom layer of agar leads to the intensive formation of colonies of myeloid cells, including cells at stages of differentiation from myeloblast to mature granulocyte. The colonies reach their largest size 12–14 days after explantation. On the average, they contain 10^3 cells. Conditioned medium from human spleen cultures also stimulates mouse bone marrow cells, whereas conditioned medium from mouse spleen cultures stimulates only mouse bone marrow and not human bone marrow cultures (Paran *et al.*, 1970).

The continuous presence of stimulating factors in the medium is necessary not only for the initial formation of colonies, but also to maintain proliferation of the cells of which they are composed. For instance, if colonies at the 4th–7th day of incubation are transferred whole to fresh nutrient medium containing 0.5 ml leukemic mouse serum to 1 ml medium, more intensive growth than before the transfer is observed. If the colony is transferred to medium containing the same quantity of healthy mouse serum, the colony decreases in size (Metcalf and Foster, 1967). If dissociated cells from 3–7 day colonies are transferred to fresh nutrient medium containing leukemic serum, macrophage growth is observed for some time. Some of the monocytes give small colonies that are formed through further proliferation of the dissociated cells. Differentiation to granulocytes ceases after subculture (Metcalf and Moore, 1971).

The number of colonies in agar bone marrow cultures correlates with the number of stem cells in the original suspension (Wu *et al.*, 1968). This was shown in parallel experiments in which suspensions of hematopoietic cells were explanted *in vitro* and transplanted into irradiated mice. The number of colonies in the culture was counted and compared with the number of hematopoietic foci in the spleen (the model of McCulloch and Till). Hematopoietic colonies in the spleen are known to be clones; i.e., each colony arises from one ancestral cell. This has also been shown to be true for colonies in cultures by chromosome analysis of the metaphase plates found in the colonies. All metaphase plates in one colony have been shown to carry a distinctive chromosomal anomaly—i.e., they contain a common marker chromosome—although the original suspension used for the culture was a mixed population in which only some of the cells contained the marker.

Later work showed that the cells responsible for colony formation in agar are committed myeloid precursors, not stem cells (Metcalf and Moore, 1971). These precursors differ from hematopoietic stem cells in several properties;

sedimentation during gradient centrifugation, sensitivity to vinblastin, changes in number after addition of an adjuvant, degree of self-support, etc. (Haskill *et al.,* 1970; Moore *et al.,* 1970). With the agar cloning method, the number of myeloid precursors in various populations of hematopoietic cells can be estimated quantitatively, and changes in their number in hematopoietic tissue can be studied during exposure to factors that disturb the equilibrium of that tissue.

Besides colonies of myeloid cells and macrophages, the formation of megakaryocytes capable of proliferating *in vitro* is also observed under certain conditions in agar cultures of bone marrow cells. Mouse bone marrow and spleen cells formed pure or mixed colonies of up to 80 megakaryocytes in agar after stimulation by medium conditioned by activated lymphoid cells (Metcalf *et al.,* 1975). Lymphoid cells produced the factor stimulating megakaryocyte proliferation after culture in medium containing 2-mercaptoethanol, with or without added mitogens.

Even if erythropoietin is added to agar cultures, erythroid colonies cannot be obtained. These colonies do arise, however, when embryonic liver cells are cultured in plasma gel and, in addition to the ordinary conditioned medium (from renal tubule cultures), erythropoietin is also added (Stephenson *et al.,* 1971). Both myeloid and erythroid colonies are formed under these circumstances. If the ordinary conditioned medium was left out, only erythroid colonies developed, but they did not increase in number. Conversely, if the cultures were grown with conditioned medium but without erythropoietin, only myeloid colonies were formed, and they likewise did not increase in number. Early erythropoetin-responsive red cell progenitors can be detected in agar cultures if thiol and high doses of erythropoetin are added to the medium (Iscove and Siber, 1975). By the 10th day, some of the erythroid colonies in such cultures increase in size to macroscopic dimensions (10^4 cells).

5. Organ Cultures of Hematopoietic Tissue

Embryonic Liver

The results of organ cultivation of the embryonic liver depend largely on the composition of the nutrient medium and the technique of explantation used.

In agar cultures of human embryonic liver, the hematopoietic cells start to die on the 4th day after explantation; at the same time, complex parenchymatous structures such as bile ducts and cysts develop in the explants (Hillis and Bang, 1962). When the method of Wolff and Haffen was used, hematopoiesis was maintained for about 10 days in mouse embryonic liver cultures (Gallien-Lartique, 1966).

Under these conditions, erythroid hematopoiesis ceases a few days after explantation. Myeloid forms are produced in the cultures even in the absence of erythropoiesis. Erythropoiesis can be activated by the addition of plasma from anemic animals to the nutrient medium (Gallien-Lartique, 1966). Mitoses are observed in foci of erythropoiesis in these cultures for 5 days, after which erythropoiesis gradually disappears.

By means of the author's suggested modification of the method of organ culture on millipore filters, not only differentiation of parenchymatous tissue, but also prolonged hematopoiesis, can be obtained *in vitro*. Fragments of liver from mouse embryos were cultured on AUFS millipore filters (0.6–0.9 μm) on medium 199 containing bovine serum, chick embryo extract, glucose, vitamin C, L-glutamine, and sodium β-glycerophosphate (Luriya *et al.*, 1969*a–c*).

In 3-day cultures of liver fragments from 16–20 day embryos, the tissue in the center of the explant degenerates, but the epithelial layers and the myeloid and erythroid cells and megakaryocytes in the peripheral zones remain intact. Growth of epithelial membranes with colonizing hematopoietic cells takes place around the circumference of the filter. Growth of histiocytes is observed on the lower surface of the filter.

On the 6th day, the fragments are surrounded by wide epithelial membranes with numerous cells in a state of mitosis. The peripheral zone of the explant is occupied by extensive areas of hematopoiesis containing cells of the myeloid, erythroid, and megakaryocytic series, distributed among the epithelial cells.

After 8 or 9 days, the epithelial membranes growing on the filter become stratified. The epithelium in them acquires the characteristic morphology of polygonal hepatocytes grouped in columns. Numerous erythroid and myeloid cells forming large, confluent foci are arranged on the membrane. Hematopoietic cells at different levels of maturity and cells in a state of mitosis are found in the focus. In the central zone of the explant, regeneration of the hepatocytes takes place with the formation of bile ducts.

On the 12th or 17th day of culture, the explant becomes flatter, and its connective tissue cells penetrate deeper into the pores of the filter. The layer of hepatic epithelium formed by large cells with clearly defined borders and polygonal in shape spreads over the smaller branching connective tissue cells to form a network penetrating into the pores of the filter and spreading over its lower surface. At this time proliferation of epithelial buds, cysts, and bile ducts with swellings on their ends takes place in some parts of the explant. Intensive hematopoiesis is observed in cultures at this time (Fig. 13). Myeloid cells in different stages of differentiation and megakaryocytes are arranged in extensive zones above the epithelial layer, and they form the top layer of the explant. Groups consisting of 4–10 megakaryocytes, lying close together, are a conspicuous feature (Fig. 14). The zones of hematopoiesis, frequently composed of several thousand cells, have no clearly defined boundaries; one zone merges with another. Nevertheless, within the zone, smaller foci of hematopoiesis can be

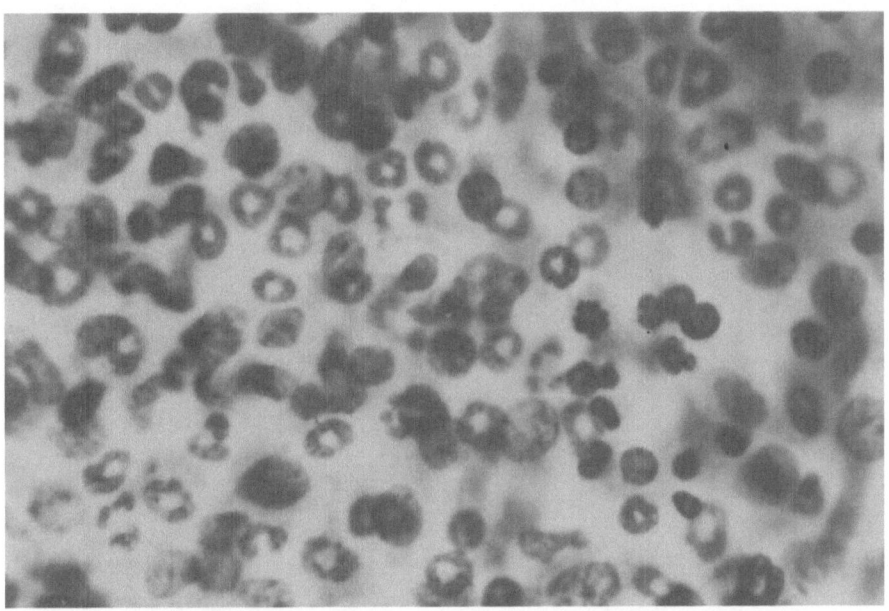

Fig. 13. Myeloid hematopoiesis in organ culture of mouse embryonic liver. Age of culture, 12 days.

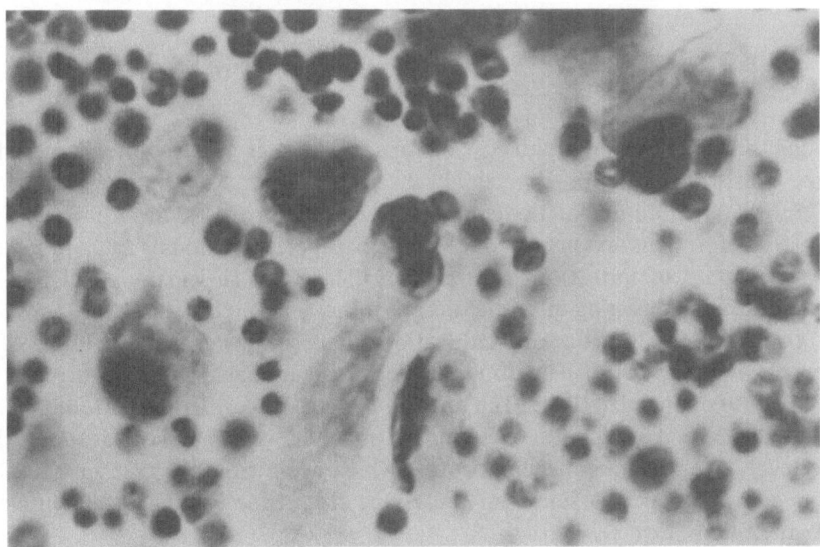

Fig. 14. Megakaryocytes in 15-day culture of embryonic liver.

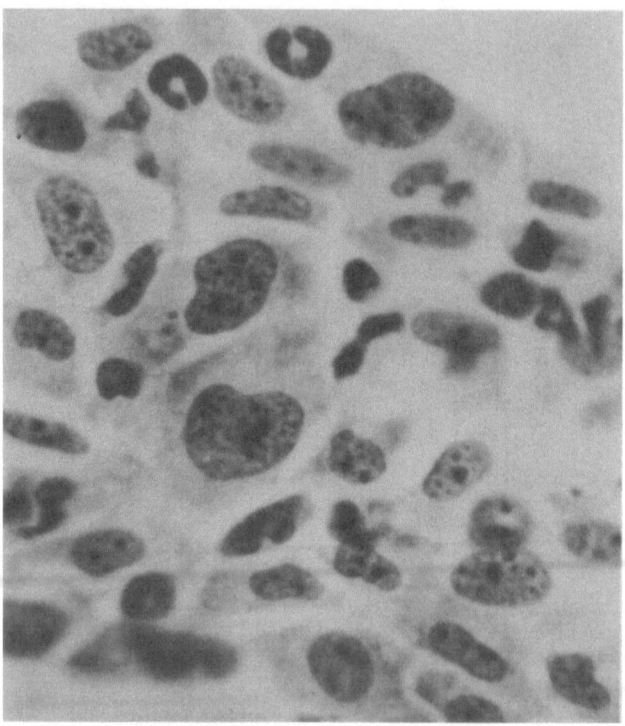

Fig. 15. Section through 20-day culture of embryonic liver.

distinguished by their morphology and arrangement as substructures, especially those consisting of groups of myeloid cells at the same stage of maturation.

Further maturation of the epithelium is visible in the 20-day cultures (Fig. 15). As before, the extensive zones of hematopoiesis including hematopoietic cells in a state of mitosis are still present. There are also many small foci of hematopoiesis, consisting of several dozens of hematopoietic cells.

On the 24th–34th days of cultivation, there are fewer hematopoietic cells than previously. Foci of hematopoiesis lie among epithelial columns surrounded by bands of connective tissue. The number of cells in the foci varies from tens to hundreds (Fig. 16). Zones of maturing myeloid cells are visible above the epithelial layer. Numerous monocytes, which were not observed previously, begin to appear at this time and the connective tissue cells spread intensively over the filter.

It was later shown that with the use of Parker 199 medium (Microbiological Associates, Inc.), more intensive and prolonged hematopoiesis can be obtained:

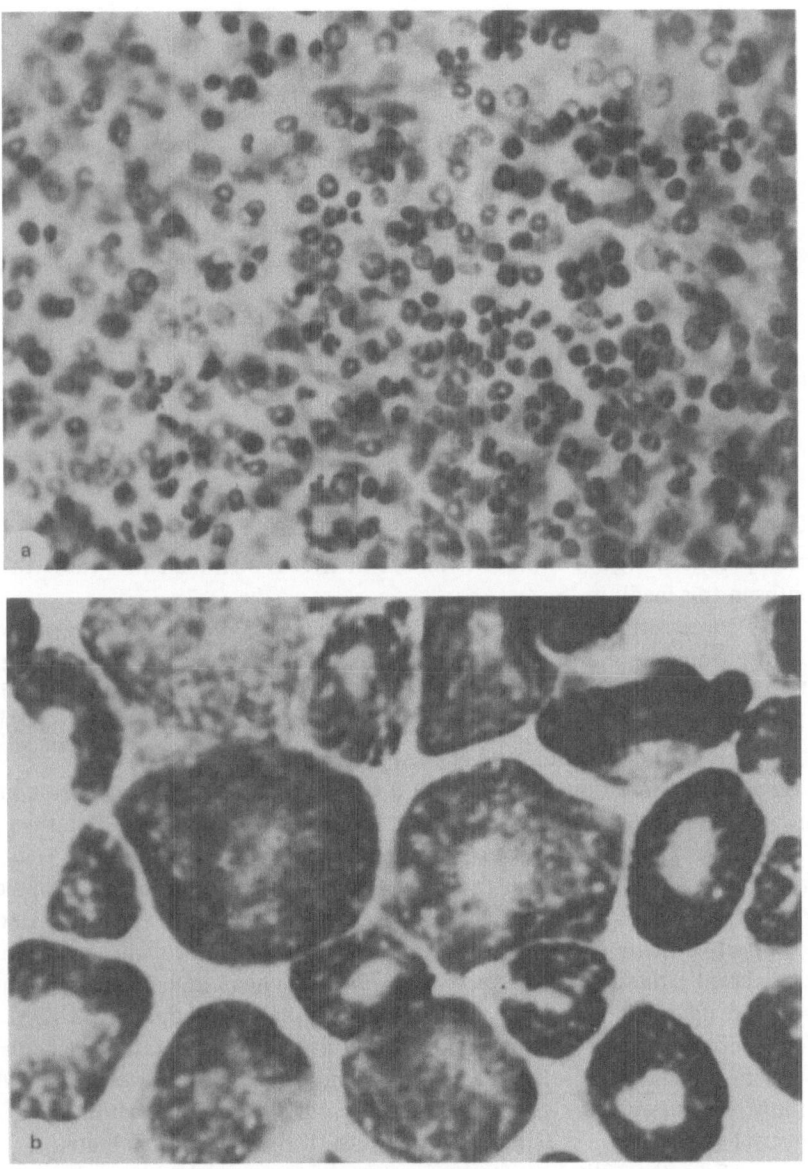

Fig. 16. Hematopoiesis in 24-day culture of embryonic liver.

TABLE 3. Composition of Hematopoietic Tissue (in %) in Organ Cultures of Liver of 17-Day Mouse Embryo[a]

Age of culture (days)	Hemocytoblasts	Myeloblasts	Promyelocytes	Myelocytes	Metamyelocytes	Polymorphs	Erythrocytes	Normoblasts		
								Basophilic	Polychromatophilic	Oxyphilic
0	7	2	1	1	–	–	10	11	22	46
5	1	5	11	15	23	27	8	8	1	1
10	5	6	19	18	31	18	1	1	1	–
14	1	6	24	16	22	31	–	–	–	–
24	2	6	24	26	30	12	–	–	–	–

[a]After Latsinik *et al.* (1969).

hematopoiesis in organ cultures of embryonic liver is maintained for more than 2 months in this medium.

The composition of the hematopoietic cells was determined from smears prepared from the cultures aged from 5 to 24 days (Table 3). It is clear from Table 3 that until the 5th day after explantation, both erythroid and myeloid series of hematopoiesis continue, but later only the myeloid series (including megakaryocytes) can be observed (Latsinik *et al.*, 1969).

The age of the embryo from which the liver is taken from explantation has a marked effect on the character of hematopoiesis in the cultures (Latsinik *et al.*, 1970*b*). In explants of the liver from 13-day embryos, hematopoietic cells are not found in most cultures on the 16th–18th days. Cessation of hematopoiesis correlates with replacement of the liver parenchyma (except the cysts and ducts) by connective tissue cells. Liver explants from 17-day mouse embryos, in which the epithelial structures are preserved and intensive hematopoiesis is maintained for 24–26 days, are optimal for the maintenance of hematopoiesis *in vitro*. In cultures of the livers of newborn mice, the intensity and duration of hematopoiesis are somewhat less. Small foci of myeloid cells persist, however, in 24-day cultures from newborn mice. Gradual cessation of erythropoiesis during the first days after explantation is observed in cultures of the livers from all donors used (embryos from the 13th day of gestation to newborn animals).

Hematopoietic cells proliferate actively in embryonic liver cultures, as shown not only by the high mitotic index, but also by the intensive incorporation of thymidine-H^3 (Fig. 17) (Luriya *et al.*, 1971*b*).

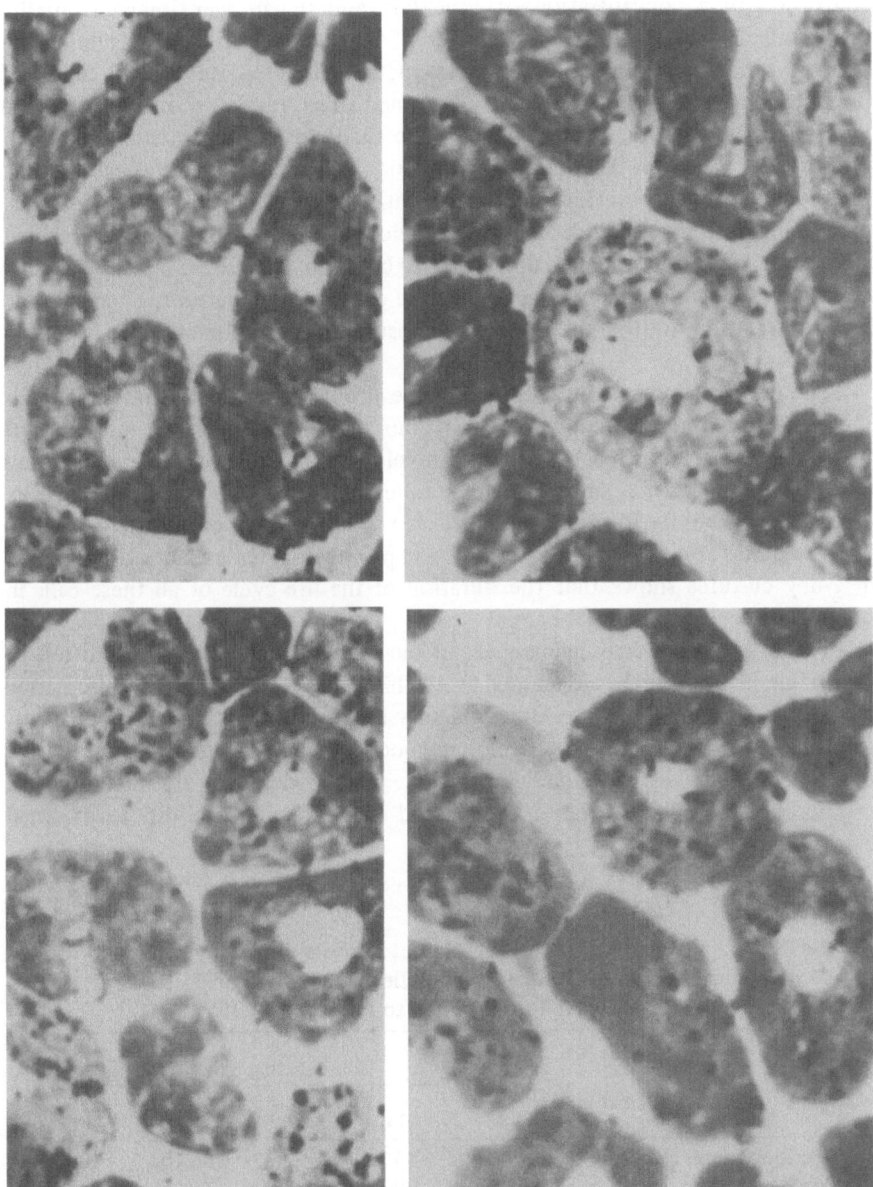

Fig. 17. Incorporation of thymidine-H³ into myeloid cells in 14-day culture of embryonic liver.

Thymidine-H^3 was added to the nutrient medium on the 4th–18th day of cultivation in a concentration of 1 μCi/ml, and smears were taken from the cultures 1–24 hr later. The percentages of labeled and unlabeled cells of the myeloid series—myeloblasts–myelocytes, metamyelocytes, stab (staff) cells, and polymorphs—in the smears were determined. The results showed that from the 5th to the 19th day, the hematopoietic tissue in the explants exhibits high proliferative activity. During growth for 24 hr in medium containing thymidine-H^3, incorporation of the label takes place into 100% of myeloblasts, promyelocytes, myelocytes, and metamyelocytes. The proportion of labeled stab cells is 71–79% labeled polymorphs, between 50 and 73%. The labeling indices for the various categories of myeloid cells in 7-day cultures of embryonic liver in relation to the duration of exposure to thymidine-H^3 are given in Table 4.

The metamyelocytes are included in the proliferating pool. That they are is shown by the finding that in the first hour of cultivation in the presence of thymidine-H^3, many (30%) of the metamyelocytes are labeled. However, not all mitoses found in the smears at this time contained the label. Metamyelocytes with the ability to proliferate were also found in older cultures.

The observation that all myeloblasts and promyelocytes are labeled after 19 hr in 7-day cultures shows that the duration of the life cycle of all these cells in organ cultures is less than 19 hr.

The first labeled stab cells appear in the cultures 7 hr after the addition of thymidine-H^3. It can be concluded from this finding that the maturation time from metamyelocyte to stab cell is evidently about 7 hr. By the same arguments, the time for differentiation from metamyelocyte to polymorph is more than 7.5 but less than 19 hr. On the whole, these results demonstrate the intensive proliferation and differentiation of myeloid cells under the conditions of organ culture.

TABLE 4. Formula of Myeloid Hematopoiesis in 7-Day Cultures of Embryonic Liver and Changes in Labeling Index 1, 3, 7.5, 19, and 24 Hours after Addition of Thymidine-H^3 to Nutrient Medium[a]

Myeloid cells	Formula (%)	Labeling index (%)				
		1 hr	3 hr	7.5 hr	19 hr	24 hr
Myeloblasts–myelocytes	35	54	65	91	100	100
Metamyelocytes	13	30	53	73	90	100
Stab cells	40	0	0	7	64	79
Polymorphs	12	0	0	0	24	54

[a]After Luriya *et al.* (1971*b*).

In organ cultures, the hematopoietic tissue of embryonic liver can thus continue to proliferate and differentiate for a long time. Hematopoiesis is probably maintained in the cultures by contact between the hematopoietic tissue and the embryonic hepatic parenchyma. There is reason to suppose that the hepatic stroma exerts specific histogenetic action on hematopoietic cells. However, no data to indicate the mechanism of these effects are yet available.

Correlation between synthesis of a specific serum protein of the hepatocytes and the duration of hematopoiesis was established by the study of embryonic liver cultures (Luriya *et al.*, 1969a,b). The protein in question is the embryonic serum protein α-fetoprotein, which is synthesized and secreted by the parenchymatous cells of the liver. There is no direct evidence that the secretion of α-fetoprotein is a factor in the regulating of embryonic hematopoiesis, but there is evidence of a close link between these two processes.

To determine α-fetoprotein, a micromethod based on double diffusion in gel with a test system for α-fetoprotein was used (Abelev *et al.*, 1963). This test system consists of rabbit antiserum against neonatal mouse serum, exhausted with adult mouse plasma, with the addition of neonatal mouse serum. The strength of the reaction was estimated by comparing the precipitation bands obtained with a standard scale of double dilutions of antigen. Serum albumin in the medium was determined similarly. The test serum consisted of rabbit antiserum against adult mouse serum together with purified albumin in optimal dilution.

Incorporation of labeled amino acids in α-fetoprotein and albumin in the cultures was also determined by immunoautoradiography (Abelev and Bakirov, 1967). A solution of glycine-C^{14} in a concentration of 2 μCi/ml was added to the culture medium. After incubation for 2 days with the label, the culture medium was concentrated, and the agar diffusion test was then performed in the usual way. The agar plates were washed, dried, and exposed on photographic films. Blackening of the autoradiograph by the labeled precipitate indicated incorporation of label into the antigen. The results of determination of α-fetoprotein and albumin during growth of the cultures showed that they synthesize these proteins over a period of 24 days *in vitro* (Table 5). The degree of synthesis correlates with hematopoiesis in the cultures, and is largely determined by the composition of the culture medium. Addition of L-glutamine to the medium, for instance, not only promotes proliferation and maintenance of differentiation of the hematopoietic cells, but also leads to more intensive and prolonged synthesis of α-fetoprotein and albumin (Table 5).

The existence of true synthesis of α-fetoprotein is confirmed immunoautoradiographically—from the incorporation of labeled amino acid into the test antigens.

By the use of a method based on a combination of electrophoresis in polyacrylamide with direct autoradiography (Abelev, 1973), the complete spec-

TABLE 5. Serum Albumin (A) and Embryonic α-Fetoprotein (α-f) in Organ Cultures of Mouse Embryonic Liver[a][b]

| Culture | Antigens | Duration of cultivation (days) | | | | | | | | |
| | | Without glutamine | | | | | | With glutamine | | |
		3	5	7	9	12	15	19	23	25
Liver of	A	+++	++	+						
18-day embryos	α-f	+++	+++	+	+	+	+			
	A	+++	++							
	α-f	+++	+++							
	A	+++		+++	+++	+++	++			
	α-f	+++		+++	++	+				
	A							+++	+++	+
	α-f							+++	++	+
Adult bone marrow	A	–		–	–	–	–			
	α-f	–		–	–	–	–			
Embryonic bone	A	–	–	–	–	–	–			
	α-f	–	–	–	–	–	–			

[a]After Luriya *et al.* (1969*a*).
[b]Legend: (+++) Distinct positive reaction for α-fetoprotein (albumin)—concentration of antigen in culture medium equal to its concentration in test system; (++) concentration of antigen lower than in test system; (+) weak positive reaction; (–) no antigen detected in medium.

trum of protein synthesis has been determined in embryonic liver cultures. A mixture of C^{14}-labeled amino acids was added to the culture media for 2 or 3 days. After removal of the free label by dialysis, the media were then subjected to electrophoresis. Proteins of the bovine serum contained in the nutrient medium and proteins synthesized by the cultures were observed on the gel after electrophoresis. To obtain the autoradiograph, the polyacrylamide gel was cut into strips, dried, and coated with fluorographic film. After exposure, impressions of the radioactive zone were obtained on the film. To identify the proteins detected on the autoradiographs, electrophoresis were carried out on media previously exhausted by antisera against albumin, α-fetoprotein, and transferrin. It was shown that albumin, α-fetoprotein, transferrin, as well as another four or five as yet unidentified components (two of which have higher electrophoretic mobility and migrate between α-fetoprotein and transferrin, while two or three are slower-migrating proteins migrating before transferrin), are synthesized in liver cultures from embryos of all ages studied (14, 16, 17, and 19 days) and from newborn mice (Fig. 18). By photometry of the autoradiographs, the ratio between the proteins synthesized can be determined at various stages of cultivation.

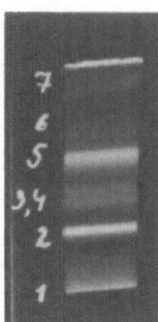

Fig. 18. Spectrum of protein synthesis in a 15-day culture of embryonic liver: (1) albumin; (2) α-fetoprotein; (5) transferrin; (3,4,6,7) unidentified proteins.

Analysis of the change in profile of protein synthesis during cultivation of embryonic liver showed that in the late stages of cultivation, the intensity of synthesis of α-fetoprotein falls, while the active synthesis of albumin and transferrin, proteins characteristic of the blood serum in adult animals, is maintained.

A correlation was observed between hematopoiesis and the synthesis of α-fetoprotein in the cultures, extinction of synthesis of α-fetoprotein and cessation of hematopoiesis take place at roughly the same time, when the active synthesis of adult serum proteins is still observed.

If the cultures are grown in medium with 20% embryonic calf serum instead of bovine serum, the synthesis of all three serum proteins, and in particular of α-fetoprotein, is extinguished more rapidly; the intensity of synthesis of α-fetoprotein falls appreciably as early as the 9th day of cultivation, whereas synthesis of the remaining proteins ceases after the 16th day of cultivation (Table 6). Characteristic differences are also observed in the morphology of these cultures, proliferation of bile ducts and cysts lined by cubical epithelium takes place, the zone of growth of these cultures is somewhat smaller than during cultivation of bovine serum, there is no proliferation of typical sheets of epithelial cells, and hematopoiesis is found only in the early periods. Fetal calf serum is known to contain embryo-specific α-globulins. Presumably, it is these α-globulins that have the inhibitory effect on the cultures. As yet, however, no direct evidence is available on the nature of the factor in fetal calf serum that inhibits differentiation and synthetic activity of the hepatic epithelium.

On the other hand, these results indicate correlation between the times of maintenance of hematopoiesis in embryonic liver cultures with the duration of synthesis of α-fetoprotein in culture.

One gets the impression that synthesis of embryo-specific serum protein in the culture reflects the stimulant action of the liver parenchyma on hematopoiesis. However, embryo-specific proteins of heterologous serum added to the

TABLE 6. Serum Albumin (A), Embryonic α-Fetoprotein (α-f), and Transferrin (T) in Organ Cultures of the Liver of 19-Day Mouse Embryos Explanted in Medium Containing Bovine and Fetal Calf Serum[a]

Serum	Antigens	Time of cultivation (days)					
		2	5	9	13	16	19
Bovine	A	+++	+++	+++	+++	+++	+++
	α-f	+++	+++	+++	++	++	++
	T	+++	+++	+++	+++	+++	+++
Calf	A	+++	+++	+++	++	++	−
	α-f	+++	+++	++	+	+	−
	T	+++	+++	+++	++	++	−

[a]Legend: As in Table 5.

culture medium, inhibit differentiation and synthetic activity of the embryonic hepatic epithelium, and thereby terminate its role in the creation of the micro-environment for embryonic hematopoiesis in the culture.

Albumin is known to be synthesized by the liver throughout life, while α-fetoprotein is synthesized only during embryogenesis. In mice and rats, a certain quantity of α-fetoprotein is found in the serum during the first weeks after birth, after which it disappears completely. Hematopoiesis in the mouse and rat liver is a feature of the second half of intrauterine development, and is also observed during the first few days after birth. Presumably, therefore, albumin does not affect embryonic hematopoiesis, whereas the synthesis of α-fetoprotein is linked with this process.

The use of the fluorescent antibody method has shown that α-fetoprotein is actually synthesized in the epithelial cells in cultures of embryonic liver. Fluorescence is not observed over the whole of the hepatic parenchyma, but only in individual trabeculae (Fig. 19). This may indicate either the heterogeneity of the population of the hepatic epithelium or the asynchronism of the synthetic process.

It is interesting to note that synthesis of α-fetoprotein has been found to occur not only in embryonic liver cells, but also in the yolk sac of rat embryos, which is known to be a site of embryonic hematopoiesis (Gitlin *et al.*, 1964). It is possible that the synthesis of α-fetoprotein is directly related to the influence exerted by the embryonic hepatic parenchyma on the hematopoietic tissue that is responsible for the prolonged maintenance of myeloid hematopoiesis in organ cultures of embryonic liver. The other embryonic tissues do not synthesize this protein.

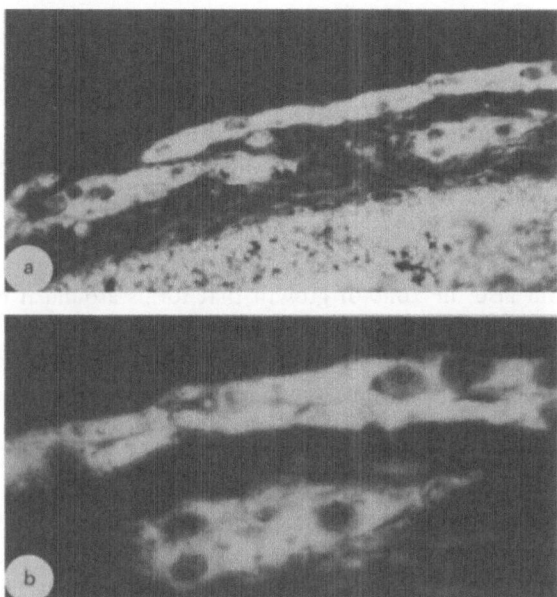

Fig. 19. Localization of α-fetoprotein in sections through 10-day culture of embryonic liver.

It can thus be postulated that the hepatic epithelium in organ cultures of the embryonic liver creates a specific microenvironment that maintains hematopoiesis. The extent to which the hematopoietic cells of the embryonic liver are sensitive to the microenvironment created by the influence of osteogenic tissue, i.e., characteristic of the adult bone marrow, is an interesting problem. When liver fragments from 17-day mouse embryos were grafted onto a previously grown osteogenic stroma (explants of the femora of 18-day embryos), intensive hematopoiesis was observed on the 7th–12th days of combined cultivation in the liver explant itself, and extensive areas of hematopoietic cells spread to the osteogenic tissue surrounding the bone. Hematopoietic cells located on the osteogenic stroma frequently form stratified foci with high proliferative activity. Hematopoiesis on an osteogenic stroma also continued at a high level in the subsequent periods of cultivation. Hence, it follows that the hematopoietic cells of embryonic liver are able to migrate in culture to osteogenic tissue, where they actively proliferate while keeping to the myeloid direction of differentiation. The problem of the extent to which the osteogenic stroma *in vitro* provides a microenvironment suitable for embryonic hematopoietic cells can evidently be solved by subculturing hematopoietic cells from embryonic liver, after successive periods of explanation in organ cultures, onto new filters with previously grown osteogenic stroma.

Bone Marrow

According to the available evidence, the fate of bone marrow in organ cultures depends largely on the species of animal from which the tissue is explanted and the structure of the bone from which the marrow is taken.

In organ cultures of marrow from the long bones of adult mice prepared on AUFS millipore filters by the multiple organ culture method, hematopoiesis continues for no more than 5 days; proliferation of the stromal cells is observed later. If millipore filters are used as the support for the explants, not only the explant itself, but also the zone of growth that forms around it can be studied (Prusevich and Luriya, 1969).

On the 5th day, the explants become flattened, and a wide zone of growth consisting of phagocytic macrophages and histiocytes forms around them on the filter. Numerous myeloid cells and megakaryocytes as well as many fat cells still remain in the fragment. Foci consisting of tens of myeloid cells can also be found in the zone of growth above cells of connective tissue type.

By the 8th day, the enlarged zone of growth consists of histiocytes and fibroblasts that are distributed haphazardly and are larger at the edges of the zone of growth. Small histiocytes with a compact nucleus and with thin, branching processes penetrate into the pores of the filter. No foci of myeloid hematopoiesis can yet be found in these cultures, but there are many leukocytes with pyknotic nuclei.

Bands of oriented fibroblasts, more closely packed than the surrounding cells, are formed at the site of the explanted fragment in the center of the zone of growth. These structures give a positive reaction for alkaline phosphatase. Foci consisting of round macrophages and histiocytes, the cytoplasm of which is filled with large pigment granules, are also present.

On the following days, the cell density in the cultures increases and debris is removed from the filters. On the 14th–18th day, blackened zones of phosphatase-positive tissue are found in most of the cultures, where they occupy a considerable area of the filter surface. They contain chiefly large fibroblasts with large nuclei containing several nucleoli and polygonal outlining of the cytoplasm. These cells are similar to osteoblasts in their morphology.

In 18-day cultures, phosphatase-positive foci consisting of closely packed polygonal cells are found. Sometimes there is a granular ground substance with numerous metachromatic granules surrounding them.

Hematopoiesis thus continues for only a short time (5 days) in organ cultures of bone marrow fragments from long bones of adult mice. Later, phosphatase-positive foci consisting of regularly and compactly arranged cells, resembling osteoblasts, are formed in the organ cultures. In some places, ground substance is formed. However, the typical structures of bone tissue, namely, trabeculae with immured osteocytes and deposits of the characteristic ground substance, were not found in the cultures. These morphological pictures may indicate that osteogenic tissue is formed in cultures of mouse bone marrow, although osteogenesis does not go on to completion.

In explants of femoral marrow from guinea pigs, hematopoietic cells likewise remained for only a short time (about 6 days). A population of histiocytes then appeared, and on the 12th–14th day, foci of large, actively proliferating fibroblasts were found on the filter. These foci were similar in many respects to those described in monolayer cultures of guinea pig bone marrow. In contrast to fibroblasts in monolayer cultures, however, those in organ cultures exhibit stratified growth; a population of histiocytes also persists, and some of them penetrate into the pores of the filter and show the morphology of stellate cells.

In some cases, areas of phosphatase-positive osteogenic tissue appear in organ cultures of guinea pig bone marrow, but this tissue does not differentiate into bone.

In contrast to the results obtained with mouse and guinea pig bone marrow, during explantation of bone marrow from human ribs in organ cultures, intensive and prolonged hematopoiesis and typical osteogenesis were observed. Bone marrow was taken from the ribs removed during operations on hematologically healthy persons. Cultures were grown on AUFS filters in medium containing 20% human serum, chick embryo extract, glucose, L-glutamine, vitamine C, and antibiotics. During the first days, emigration and proliferation of the stromal cells, which had the morphology of fibroblasts, took place. Intensive hematopoiesis was observed in close contact with the stroma cells on the filter. Until the 5th or 6th day, both myeloid and erythroid hematopoiesis took place in the cultures, but later (until the 18th day inclusive), only myeloid and megakaryocytic hematopoiesis were observed. Up to the 15th day, hematopoietic cells in various stages of maturation constituted the greater part of the surface zones of the explant. They formed extensive, confluent foci in which all categories of myeloid cells were represented. Here and there, however, groups of cells belonging predominantly to one particular line of differentiation, such as megakaryocytes and megakaryoblasts, could be observed. The numerous mitoses as well as the thymidine-H^3 labeling are evidence of the prolonged and intensive proliferation of human myeloid tissue in organ cultures. By the 8th–10th day, the stromal cells in the zone of growth on the filters acquired distinct characteristics of osteogenic tissue: they reacted positively for alkaline phosphatase and were surrounded by ground substance. Later, bone trabeculae, small to begin with but later gradually increasing in size, were formed. Intensive osteogenesis was observed not only in the zone of growth, but also among the fragments of old bone, which degenerated rapidly in the cultures. The newly formed bone had the typical structure of woven bone tissue, while the ground substance in it was well developed, containing numerous immured osteocytes, and a layer of active osteoblasts was present (Luriya *et al.*, 1972*b*).

Foci of hematopoiesis lay closer to the surface, forming the top layer of the zone of growth, and as a rule they were associated with newly formed bone. Analysis of the formula of the hematopoietic tissue in human bone marrow cultures aged 4–18 days showed that promyelocytes (Fig. 20) were predominant in the cultures at all times, accounting for about 60–70% of the total number of myeloid cells. Distinguishing features of the neutrophilic promyelocytes were the abundance and relatively large size of the granules in the cytoplasm. Many megakaryocytes at different stages of maturation were found in impressions from the cultures. Cells of the erythroid series in the 9-day cultures were limited to polychromatic normoblasts, and their nuclei were pyknotic and resembled those found in oxyphilic normoblasts.

On addition of thymidine-H^3 to the culture medium, proliferation of the myeloid cells in the cultures was demonstrated. The proportion of myeloid

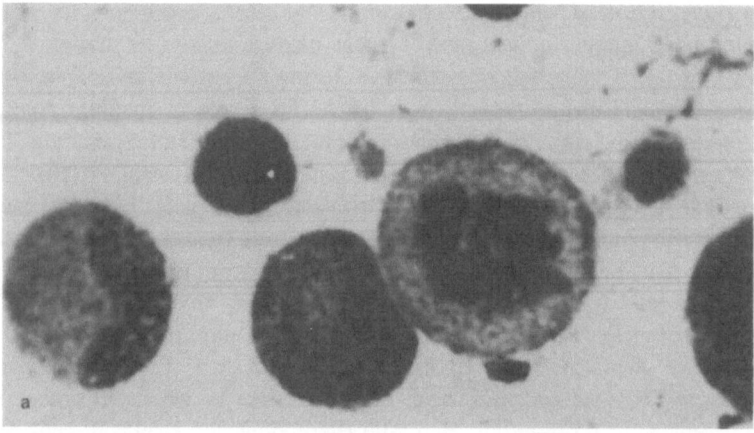

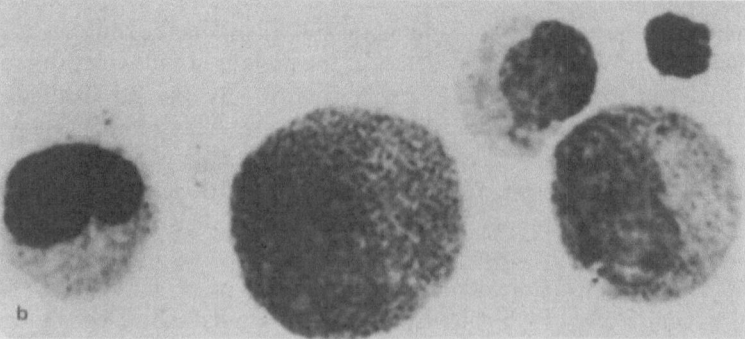

Fig. 20. Myeloid hematopoiesis in 9-day (a,b) and 14-day (c,d) organ cultures of human bone marrow.

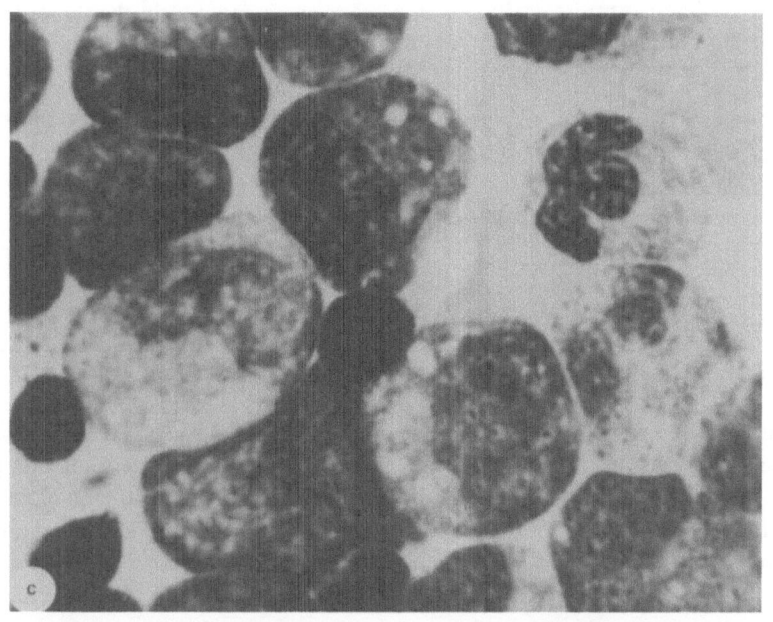

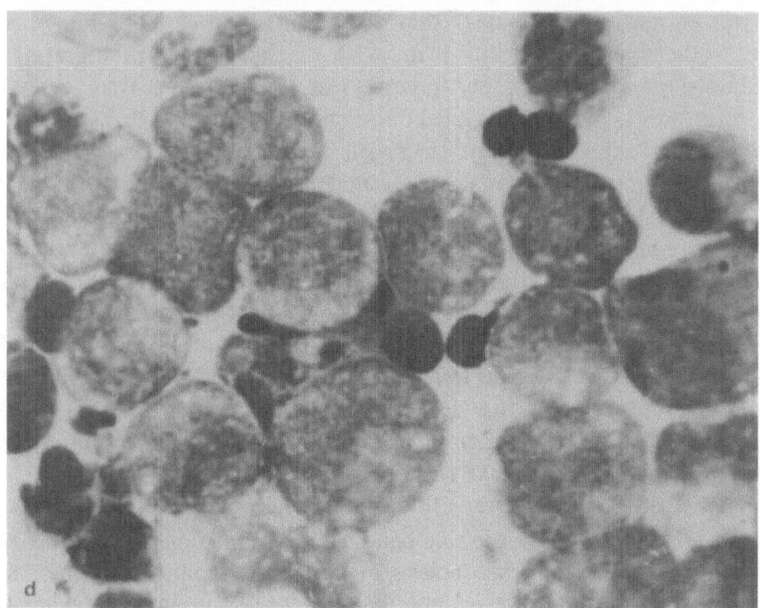

TABLE 7. Formula of Myeloid Hematopoiesis and Labeling Index in 8- to 10-Day Cultures of Human Bone Marrow Incubated in the Presence of Thymidine-H^3 for 1, 3, 9, 24, 48, and 72 Hours[a]

Categories of myeloid cells	Formula (%)	Labeling index					
		1 hr	3 hr	9 hr	24 hr	48 hr	72 hr
Promyelocytes and myelocytes	70.2	57	65	70	77	80	88
Metamyelocytes	9.7	0	25.5	37.5	45	57	64
Stab cells	9.6	0	0	20.7	35	50	59
Polymorphs	10.3	0	0	0	30	40	50

[a]After Luriya et al. (1974).

tissue cells belonging to the dividing category was not much less than normal. Those categories of myeloid cells that belong to the nonproliferating compartment in vivo do not proliferate in culture either. Maturation of the myeloid cells in culture has a number of characteristic and important differences, chief among which is the disproportionately high content of promyelocytes, which divide actively. However, the myeloid cells in such cultures go on to complete maturation.

It will be clear from Table 7 that after incubation of 8-day bone marrow cultures with thymidine-H^3 for 1 hr, 57% of cells in the proliferating pool (myeloblasts, promyelocytes, and myelocytes) are labeled. Metamyelocytes in the cultures do not take up the label during this time. After 3 hr in the presence of thymidine-H^3, the percentage of labeled metamyelocytes rises to 25.5%; no labeled stab cells can yet be found. After incubation with thymidine-H^3 for 9 hr, the index of labeled metamyelocytes rises to 37.5%, and labeled stab cells appear, with a labeling index at this time of 20.7%. Labeled polymorphs appear only after cultivation for 24 hr in the presence of thymidine-H^3. Some parameters of differentiation of cells of the myeloid series in organ cultures can be assessed from these results. For instance, the time of differentiation from myelocyte to metamyelocyte is less than 3 hr, to stab cell less than 9 hr, and to polymorph more than 9 but less than 24 hr. Among the myeloblasts– myelocytes, the proportion of proliferating cells is high, and during prolonged cultivation, it comes close to 90% (Luriya et al., 1974).

Important differences are thus found between organ cultures of mouse and guinea pig bone marrow, on the one hand, and of human marrow, on the other. The differences are that hematopoiesis in mouse and guinea pig bone marrow quickly ceases, while intensive myelopoiesis takes place in human bone marrow, at least for 2.5 weeks.

The possible cause of these differences may be either that the culture medium used is more favorable for human tissues than for mouse and guinea pig tissues, or that the bone marrow from the long bones differs from marrow obtained from cancellous bones. The second of these explanations seems most likely to be correct. The duration of hematopoiesis in cultures in fact always correlates with osteogenesis, and foci of hematopoietic cells are associated with newly formed osteogenic tissue. Endosteal osteogenic tissue is far more abundant in flat bones than in long bones.

The role of newly formed osteogenic tissue in the maintenance of hematopoiesis in organ cultures was verified directly by experiments in which bone marrow from long bones of mice was cultured on a previously grown osteogenic feeder layer (Prusevich and Luriya, 1969). To obtain the osteogenic tissue, the femoral anlagen from 18-day mouse embryos were explanted into organ cultures on a filter.

Periosteal osteogenesis was taking place in the bone explants at the time of explantation of the marrow, and an osteogenic zone of growth was formed on the filter. Most of the hematopoietic tissue was dead 6 days after explantation of the marrow. The hematopoietic cells were concentrated mainly around the fragments of bone, and also occurred as compact groups in the zone of growth. At this time, areas of completed osteogenesis, consisting of bone trabeculae with deposits of calcium were found on the filter. On the 9th and especially on the 11th day, intensification of osteogenesis and myeloid hematopoiesis took place. Numerous branching bone trabeculae appeared in the stratified zone of growth. Foci of myeloid hematopoiesis, consisting of several layers, lay on their surfaces. The surface layers of the zone of growth were filled with extensive zones of intensive myeloid hematopoiesis with numerous dividing cells. Zones of myeloid cells also spread to the periphery of the zone of growth and covered much of its area.

In cultures aged 14–16 days, the number of hematopoietic cells was slightly reduced, and the myeloid cells, in various stages of differentiation, were distributed in foci of different sizes around the explant and at the periphery of the surface layer of the zone of growth. Until the 18th day after explantation of bone marrow, viable myeloid cells could be found in the cultures.

In contrast to the gradual disappearance of hematopoiesis in cultures of the femoral marrow of mice, during cultivation of the same bone marrow on an osteogenic stroma, hematopoiesis determined by interaction between hematopoietic tissue and bone tissue is thus observed for more than 2 weeks (Fig. 21).

In other experiments, a comparative analysis was made of osteogenesis and hematopoiesis during explanation of bone marrow from the long and cancellous bones of guinea pigs. As has already been stated, hematopoiesis in a bone marrow fragment from the long bones of guinea pigs ceases 4–5 days after explanation, and no bone tissue is formed. In cultures of bone marrow from the pelvic (cancellous) bones of guinea pigs, osteogenic tissue was formed on the

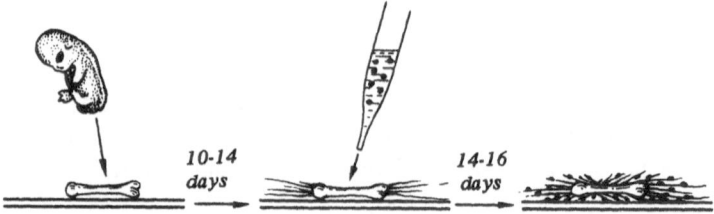

Fig. 21. Diagram of explantation of bone marrow on previously grown osteogenic stroma.

filter on the 7th–10th day. Under these conditions, hematopoiesis was observed for 12 days and more.

6. A Comparative Assessment of the Various Methods of Hematopoietic Tissue Culture

Duration of Hematopoiesis

Most methods used for explantation of hematopoietic tissue do not permit hematopoiesis to be maintained for a long time *in vitro.* For example, explantation of bone marrow in a plasma clot or in monolayer and suspension cultures gives only short survival of the hematopoietic cells; later, the hematopoietic cells are replaced by a connective tissue cell population. Explantation of embryonic liver tissue yields somewhat different results. Active hematopoiesis in embryonic liver cultures in a plasma clot has been observed for 7–10 days.

Organ culture methods enable hematopoiesis to be maintained longer, on the whole, than do monolayer, suspension, and plasma cultures. In these methods, the duration of hematopoiesis *in vitro* is determined largely by the method of organ culture used and by the tissue explanted. If pieces of human embryonic liver are explanted on agar, the hematopoietic cells degenerate after 4 days (Hillis and Bang, 1962). Culture of hematopoietic tissue on rafts likewise has not given satisfactory results (Chen, 1954). Somewhat better conditions for the maintenance of hematopoiesis are created in agar cultures prepared by the method of Wolff and Haffen (1952); hematopoiesis can be maintained for 7–10 days in explants of embryonic liver, bone marrow, and spleen (Poureau-Schneider, 1962, 1963; Salvatorelli, 1966, 1967*a,b*)

If embryonic mouse liver is cultivated on millipore filters by the multiple organ culture technique, myeloid hematopoiesis can be maintained for more than 2 months. The essential features of this method is that intensive proliferation and differentiation of hematopoietic cells, including hematopoietic stem cells, take place (see below).

If bone marrow from the long bones of adult rodents is explanted by the same method, hematopoiesis continues only for a short time (5 days). Explantation of bone marrow onto a previously grown bone stroma gives different results. Under these conditions, myeloid hematopoiesis is maintained for more than 2 weeks. Prolonged (over 2 weeks) and intensive hematopoiesis takes place in organ cultures of human bone marrow obtained from the flat bones (the ribs of adults).

So far as the duration of hematopoiesis *in vitro* is concerned, the multiple organ culture technique thus gives significantly better results than all other culture methods so far used.

Myelopoiesis and erythropoiesis in culture differ in certain respects.

Myelopoiesis

Myeloid cells remain viable for only a few days in submerged bone marrow cultures: in a plasma clot; in a Jayne and Pulvertaft chamber, in which the explant is placed in a drop of medium of nutrient agar; and also by the use of other types of culture chambers. According to Rasmussen (1933), myeloblasts in bone marrow cultures from young rabbits proliferate in the first days after explantation and are sometimes transformed into myelocytes. Maksimov (1916) found that myelocytes in plasma cultures can divide by mitosis for 5 days without progressive differentiation.

Conflicting opinions are held on the possibility of further maturation of myelocytes in submerged cultures. Lajtha (1965), for instance, considers that myelocytes in a liquid nutrient medium progress to mature polymorphs in 24 hr. Other workers consider that myelocytes do not differentiate into more mature forms in submerged cultures (Pulvertaft and Humble, 1956).

Polymorphonuclear neutrophils contained in the original material exhibit motility in various types of tissue cultures for several days (Thomas, 1956; Woodliff, 1958, 1964). Most of these cells degenerate soon after explantation. Polymorphonuclear basophils may persist up to 4 days in submerged cultures.

In organ cultures prepared by the method of Wolff and Haffen (1952), myeloid hematopoiesis continues for 10–12 days (Poureau-Schneider, 1962, 1963).

Foci consisting of myeloid cells or monocytes develop in agar cultures of a suspension of hematopoietic cells on a feeder layer, or with the use of conditioned media (Pluznik and Sachs, 1965, 1966; Bradley and Metcalf, 1966). The cells in the myeloid foci retain their specific differentiation and proliferative activity for 10–14 days.

As investigation of chromosome markers has shown, each colony is a clone; i.e., it arises from a single myeloid precursor cell. During growth of the culture, the foci of myeloid cells are replaced by foci of mononuclear cells, macrophages, and histiocytes.

Although the number of precursors of colonies growing in agar cultures correlates with the number of cells forming colonies in the spleens of irradiated mice (Wu *et al.*, 1968), the precursors of colonies in agar cultures cannot be identified with those of colonies in the spleen. Cells forming foci on agar are committed precursors to the myeloid form of differentiation, capable of supporting themselves only for a limited time (Wu *et al.*, 1968).

In all these established methods of culture, therefore, hematopoiesis takes place *in vitro* on account of precursors that have already begun to differentiate at the time of explantation. That is why the duration of myelopoiesis in culture is limited to the period of self-maintenance of committed myeloid precursor cells, while the number of young forms of the myeloid series in the explant decreases progressively with the age of the culture. Hematopoietic stem cells seem not to participate in myelopoiesis in agar culture. According to Metcalf and Moore (1971), the vast majority of 7–10 day colonies in agar cultures lack detectable hematopoietic stem cells. The number of committed myeloid precursors in these cultures is considerable only during the first few days after explantation, after which it falls sharply. Observations on myeloid colonies on agar for 10–14 days show that proliferation of the precursors does not take place in the late periods of cultivation, but that mature forms persist *in vitro*.

Explantation of hematopoietic tissue by the multiple organ culture technique yielded different results. In organ cultures of mouse embryonic liver, myelopoiesis is maintained for over 2 months if optimal conditions are provided. The study of the cell composition of these cultures shows that the myeloid cells not only mature but also undergo intensive proliferation and differentiation. According to results obtained with the use of thymidine-H^3, myeloid cells in organ cultures proliferate for a long time. The size of the proliferating pool remains very large. Meanwhile, maturation of the proliferating cells takes place in the cultures, and undividing cells of the myeloid series are formed from them.

On a feeder layer consisting of previously grown bone tissue, myeloid hematopoiesis continues for 16–18 days in bone marrow explants from the long bones of mice. In organ cultures of human bone marrow taken from flat bones, myeloid hematopoiesis is also maintained for not less than 18 days. The use of thymidine-H^3 showed that under these circumstances, myeloblasts–myelocytes proliferate, differentiate, and mature into polymorphs.

Hematopoiesis in these cultures is distinctly promyelocytic in character, with immature forms sharply predominant over mature. The partial block in the development of myeloid cells at the promyelocyte stage is probably due to the unusual conditions under which the explanted hematopoietic tissue exists. Not only the process of cell repopulation, which lies at the basis of histogenesis of hematopoietic tissue, but also the influence on this process of controlling factors originating from other organs and tissues (including the hormonal background), are eliminated during growth in culture. The primary packing of the cells characteristic of the intact bone marrow is also dispersed in marrow explants.

Cloning of megakaryocytes takes place in agar (Metcalf *et al.*, 1975). Megakaryocytes remain for only a short time in submerged bone marrow cultures, and they are capable of forming platelets (Thiery and Bessis, 1956*a,b*; Pulvertaft and Humble, 1956). In cultures of mouse and rat bone marrow 5 hr after explantation, numerous long bands appear in the cytoplasm of the megakaryocytes; these bands later break up into platelets, and the process of platelet formation is activated by a low oxygen concentration (Pulvertaft, 1958).

Megakaryocytes persist for several weeks in multiple organ cultures of human bone marrow, of mouse bone marrow on a bone stroma, and in explants of embryonic liver. They frequently form foci or colonies consisting of more than 10 cells.

Observations on prolonged myeloid and megakaryocytic hematopoiesis in multiple organ cultures have raised the question whether hematopoiesis is maintained in such cultures entirely by proliferation and maturation of committed myeloid precursors, or whether it is maintained through the participation of hematopoietic stem cells.

The time taken for myeloid cells to pass through the proliferative phase *in vivo* is known to be about 4 days in mice and 8 days in man (Bond *et al.*, 1965). Our own observations on myeloid hematopoiesis over a period of 16–18 days in mouse bone marrow grafted on a bone stroma, and in cultures of human bone marrow for 18 days, accordingly constitute substantial evidence that in organ culture, proliferation and differentiation of hematopoietic stem cells take place. This hypothesis, however, requires experimental confirmation because the long period of hematopoiesis in culture can be explained in another way: by the possibility of changes in the properties of the committed myeloid and megakaryocytic precursors in explanted hematopoietic tissue grown in multiple organ cultures, so that they become capable of dividing by mitosis more times than *in vivo* (i.e., they can maintain themselves for a longer time).

There is direct evidence, however, to show that hematopoietic stem cells not only exist, but also proliferate intensively in organ cultures (see p. 68). It is noteworthy that the life span of hematopoietic stem cells in culture correlates with the duration of maintenance of hematopoiesis. On the other hand, the parameters of differentiation and proliferation of myeloid cells in culture determined from the incorporation of thymidine-H^3 were in any event not lengthened, but were rather shortened compared with the corresponding parameters *in vivo*.

All these arguments show that myeloid hematopoiesis in multiple organ cultures differs qualitatively from hematopoiesis in other cultures. In organ cultures, there is intensive myelopoiesis, continuing significantly longer than the period of complete maturation of the youngest myeloid precursors, and a pool of immature cell forms is maintained for a long time in the population. These distinguishing features of hematopoiesis in multiple organ cultures seem to depend on the participation not only of committed myeloid precursors with a limited period of self-maintenance, but also of hematopoietic stem cells.

Erythropoiesis

Whether progressive differentiation of erythroid clones in agar can take place or not depends largely on the type of hematopoietic tissue (embryonic or adult) used for explantation and on method of culture. Erythropoetin, thiol, and colony-stimulating factor seem to be essential for erythroid colony formation.

In most investigations involving explantation of bone marrow in a plasma clot, the findings described indicate that erythroid cells migrate from the explant and subsequently degenerate during the first few days of culture. There is some evidence, however, to show that cells of the erythroid series can differentiate even under these conditions. For instance, Rachmilewitz and Rosin (1944) described maturation of erythroid cells in plasma cultures of rabbit bone marrow. In serum—agar cultures of bone marrow, mitoses can be seen among the normoblasts, and in rare cases, the nucleus is expelled from late normoblasts, although most cells degenerate without undergoing differentiation (Pulvertaft and Humble, 1956).

In plasma cultures of human embryonic liver, Benevolenskaya (1929) observed erythropoiesis that continued for 7–10 days. Mature anuclear erythrocytes persisted in the vessels of the explant for more than a week after erythropoiesis ceased. Sorieul (1966) observed that erythroid and myeloid hematopoiesis continued in the mouse embryonic liver when cultured in roller tubes up to 3 weeks. The investigations of Benevolenskaya and Sorieul, it must be noted, are the only ones in which erythropoiesis is described over a long period of time *in vitro*.

In organ cultures set up by the methods of Wolff and Haffen (1952) and Luriya *et al.* (1969*b*), erythropoiesis declines just as in most cases when submerged cultures are used, i.e., sooner than myelopoiesis. In embryonic liver cultures, erythropoiesis ceases after 5 days *in vitro*.

There is evidence that certain procedures can prolong erythropoiesis *in vitro*. For example, maintenance of erythropoiesis for about 2 weeks has been described in organ cultures of bone marrow from a chick embryo aged 18 or 19 days when explanted along with chick embryo liver, whereas in control cultures, it ceases after 3 days (Salvatorelli, 1967*a,b*). An extract and dialyzate from embryo liver also stimulated erythropoiesis in bone marrow cultures (Salvatorelli, 1967*a*). Maintenance of erythropoiesis in cultures of mouse embryonic liver has also been described after the addition of plasma from anemic animals to the culture fluid (Gallien-Lartique, 1966).

7. Factors That Influence Differentiation of Hematopoietic Tissue *In Vitro*

The behavior of hematopoietic tissue in culture is an excellent demonstration of how the differentiation of its cells depends on the action of external factors, an idea steadfastly promoted by A. A. Maksimov. After isolation of hematopoietic tissue, and especially after separation of its cells, hematopoiesis ceases. This is the situation that arises in suspension and monolayer cultures and also in the zone of growth of plasma cultures. Conversely, interaction among the cells both of the tissue itself and of neighboring tissues composing the hematopoietic organs is preserved in organ cultures. In this way, as has been pointed out, the conditions for prolonged maintenance of hematopoiesis are created in organ cultures. Local intercellular interactions (within the same or between different tissues) seem to be most important factors for differentiation of hematopoietic cells *in vitro,* and interaction with cells of the stroma is particularly important. This conclusion, drawn from our experience of work with organ cultures, is in good agreement with data showing that for differentiation of transplanted hematopoietic cells, their contact with the genetically normal stroma of hematopoietic organs is needed (McCulloch and Till, 1970), and that the type of hematopoiesis (mainly the direction of differentiation of the progeny of the hematopoietic stem cells) is determined by the microenvironment created by the stroma of the hematopoietic organ (N. S. Wolf and Trentin, 1968; Didukh and Fridenshtein, 1970). It is clear from the use of models of heterotopic transplantation (Fridenshtein *et al.*, 1968; Fridenshtein and Kuralesova, 1971) that the volume of the stroma serves additionally as the chief factor limiting the size of the hematopoietic and lymphoid tissue regenerating in the grafts. In transplants of this type, excessive overgrowth of the stroma takes place only very rarely, so that the hematopoietic and lymphoid tissues formed at the site of the grafts are similar in volume to the original grafts.

The stroma of the hematopoietic tissue in organ cultures (in contrast with monolayer cultures) likewise hardly increases in size. A factor that evidently contributes to this result is differentiation of the stroma cells, so that in organ cultures, they become a suitable feeder layer for proliferation and maturation of the hematopoietic cells. This relationship between hematopoiesis and interaction with the stroma is responsible for maintaining the hematopoietic cells in organ cultures, but on the other hand, it restricts their proliferation. All these factors must be taken into account when further improvements are made in the technique of organ culture if an attempt is made to increase the cell mass of hematopoietic tissue formed in the course of cultivation.

The epithelial tissue of embryonic liver, like bone tissue in the adult, is a natural stroma for hematopoiesis. Maintenance of differentiation of the myeloid cells on a feeder layer of kidney or embryonic cells has no direct analogy with the relationships observed *in vivo* between hematopoietic and stromal tissues. In the latter case also, however, secretion of substances stimulating hematopoiesis by the cells of the feeder layer is evidently involved. Contact with hematopoietic cells is in fact not necessary for a positive effect. The feeder cells have a colony-stimulating action even if a thin layer of agar is interposed between them and the hematopoietic cells. The stimulating factor is thus liberated from the cells and can diffuse through the layer of agar without losing its activity. That conditioned media capable of inducing the formation of colonies of hemato-poietic cells can be prepared is further evidence in support of this view.

Abundant information has now been obtained on the colony-stimulating factor (CSF). As experiments on medium-conditioning have shown, CSF can be produced by various tissues of the body (Sheridan and Stanley, 1971). CSF is found in the sera of mice with spontaneous and transplanted leukemia, in some sera of patients with leukemia, and also in healthy mouse and human urine (Stanley, 1972; Stanley and Metcalf, 1971). The study of the nature of mouse and human CSF has shown that it is a glycoprotein, migrating in the postalbu-min fraction with the α_1-globulins during electrophoresis. The molecular weight of the CSF from mouse serum is between 80,000 and 150,000, whereas that of the CSF from human urine varies from 45,000 to 60,000. It is also a glycopro-tein, and it contains sialic acid. The CSF isolated from cultures of mouse embryonic cells has similar characteristics. Lipoprotein inhibitors of CSFs are present in considerable quantities in healthy mouse and human serum (Chan *et al.*, 1971). Ether extraction of a medium containing CSF leads to removal of the CSF inhibitors.

Metcalf and Moore (1973) postulate that CSF acts in the body as a regulator of granulopoiesis and monocyte formation. It is impossible, however, to deter-mine the CSF concentration in the body acting directly on the precursor cells in the hematopoietic organs, and the CSF level in the blood does not necessarily reflect changes in the local CSF concentration, should these in fact be impor-tant. Experiments on irradiated mice have shown that the CSF level in their sera rises while the concentration of the inhibitors falls. The establishment of correlation between regeneration of myeloid and macrophage precursors in the bone marrow and a dramatic increase in the ability of the cells lining the medullary cavity to produce CSF was an important observation.

In contrast, no visible changes were observed in CSF production in regenerat-ing cells of the bone marrow or lung and spleen tissue (Chan and Metcalf, 1972). It must be noted that the cells lining the medullary cavity produce CSF not only after whole-body irradiation, but also if present in the screened femur.

The parallel trend observed between the lowering of the inhibitory level and the increase in the CSF concentration suggests that a fall in the inhibitor level leads to stimulation of CSF production (Chan and Metcalf, 1972). Addition of inhibitors to a medium in which fragments of bone cylinders are incubated in fact prevents CSF production.

The bone marrow stroma is thus evidently capable of producing a local factor that stimulates regeneration of myeloid precursors. It is not yet clear, however, which stromal cells this activity is connected with: the fibroblasts, endothelial cells, macrophages, or osteoblasts. In particular, it is essential to find out whether stromal cells of different origin, growing in cultures in the form of diploid strains, can produce CSF.

According to Reisner (1966), the addition of steroid hormones to the culture medium has a significant effect on hematopoiesis *in vitro*. Estrone, for instance, stimulates granulopoiesis in human bone marrow explants. Granulocytes are arranged in groups around one or more reticulum cells. They incorporate tritiated thymidine, whereas the reticulum cell in the center remains unlabeled. Growth of fibroblasts is less intensive in treated cultures than in controls.

Under the influence of testosterone, fibroblast formation is largely inhibited. Of the stromal elements, mainly the endothelial cells proliferate. Foci of erythropoiesis are clearly detectable in these cultures, and the number of free erythroid forms containing hemoglobin increases. Incorporation of the thymidine label into proliferating cells of the erythroid series is observed.

Of all the humoral factors that affect hematopoiesis, the most information is available on erythropoietin, a hormone that stimulates erythropoiesis.

The mechanism of action of erythropoietin is considered to be analogous to derepression of the genome, as a result of which a new messenger RNA is synthesized (Krantz and Goldwasser, 1965).

Addition of erythropoietin to the culture medium leads to activation of erythropoiesis *in vitro* (Miura *et al.,* 1968). In these experiments, fragments of spleen from polycythemic mice were grown in test tubes in the presence of 0.5 U/ml erythropoietin. Synthesis of heme was determined from the incorporation of Fe^{59}. On explantation of the spleen without erythropoietin, neither erythroblasts nor incorporation of Fe^{59} could be found. Incubation with erythropoietin for 12 hr stimulated erythroblast formation and also the synthesis of heme, the concentration of which was 70% of that found in cultures incubated with erythropoietin for 48 hr.

Differentiation of erythroblasts and the synthesis of heme were completely blocked by the addition to the medium of actinomycin D in a concentration of 0.01 μg/ml.

A gradual decrease in the level of heme synthesis is observed in cultures of a suspension of human bone marrow in medium NCTC-109 with the addition of 20% human serum and 20% calf serum. After culture for 100 hr, the rate of heme formation is less than 10% of its initial value (Krantz, 1968). A different dynamics of heme synthesis has been described for cultures to which erythropoietin was added in a dose of 0.3 U/ml medium. In this case, after cultivation for 20 hr, heme synthesis was increased by 30% over its initial level. Later, the rate of synthesis fell gradually to 70% of the initial level, where it continued for a short time.

If a suspension of bone marrow from patients with polycythemia vera was used for explantation and the cultures were grown in the presence of serum from the same patient, erythropoietin had only a weak stimulant action. It must be asked whether the serum of these patients contains an inhibitor of erythropoietin or whether the hematopoietic cells themselves are modified in this disease and become insensitive to the action of the hormone. Bone marrow from normal donors, if cultured in a medium with the addition of 20% serum from these patients, responds to addition of erythropoietin to the medium by increased heme production, just as it does during cultivation on serum from healthy persons. Consequently, the second hypothesis is valid. The observations of Reisner (1967) that erythropoiesis is stimulated by the action of serum from patients with polycythemia vera were not confirmed by this investigation.

Addition of erythropoietin-active serum to the nutrient medium of organ cultures of embryonic liver did not prolong the duration of erythropoiesis. On the 6th–8th day after explantation, numerous erythrocytes with remnants of nuclei appeared in the cultures, possibly indicating acceleration of differentiation of mature cells of the erythroid series.

Analysis of the factors that influence hematopoiesis in culture is essentially only just beginning. Many humoral factors and many different types of interactions among cells, especially interactions between hematopoietic and stromal cells, still require study. These investigations must not only lead to progress in the technique of explantation of hematopoietic tissue, but must also broaden our understanding of the control mechanisms acting on hematopoietic tissue *in vivo.*

CHAPTER II

Possibilities of Differentiation of Cells Grown in Hematopoietic Tissue Cultures on Retransplantation *In Vivo*

Growth in culture of cell material suitable for transplantation may become a reasonable task in hematopoietic tissue culture. As was stated in the Introduction, it is a task that by no means applies to all tissues, hematopoietic and lymphoid tissues being exceptions. Two tissue components can be distinguished in every hematopoietic and lymphoid organ: hematopoietic (or lymphoid) and stromal. With regard to the first component, numerous experiments have now shown conclusively that its origin is in a hematopoietic stem cell that is common to every hematopoietic tissue (see Chapter V). Thus, the question of the suitability of cultured cells for successful transplantation *in vivo* can be reduced quite simply to whether hematopoietic stem cells can be maintained *in vitro*, or, to state the question as a practical problem, to the development of methods of culture that enable hematopoietic stem cells to proliferate.

There seems to be a single stem cell for differentiation into both hematopoietic and lymphoid series. The direction in which it differentiates is probably determined, not by one factor, but by several factors, including the stroma on which the stem cell rests.

The morphology of the stem cell has not yet been elucidated. The results of fractionation of bone marrow indicate that the stem cell may have the morphology of a lymphocyte. It must be noted, however, that the lymphocytes of bone marrow are a category of cell that evidently differs significantly from the lymphocytes of lymphoid organs, which they resemble only in their morphological characteristics. Other marrow cells with a long life cycle may perhaps also act as hematopoietic stem cells. One also cannot rule out the possibility that the stem cell may exist in various morphological states; i.e., it may be transformed from lymphocyte into other cell types or vice versa (Fridenshtein and Chertkov, 1969; Yoffey, 1973).

So far as the stromal elements are concerned, the task of growing them in culture as a material suitable for transplantation is also one of some urgency. Many clinical observations, experimental results, and data obtained in tissue culture (see Chapters I and III) have shown that contact with elements of the stroma is essential for the normal differentiation of hematopoietic cells. Diseases are known that are due to a disturbance of the stromal components of lymphoid and hematopoietic tissue. Transplantation of stromal tissue can provide an additional site for hematopoiesis and lymphopoiesis in the body. For this reason, besides attempting to solve the problem of growing hematopoietic stem cells in culture, it is also essential to undertake the task of growing stromal cells, while preserving the specific qualities of the stroma of the corresponding hematopoietic and lymphoid organs and assuming responsibility for maintaining their microenvironment. Investigations of both these problems will be discussed below.

Intravenous transplantation into irradiated recipients provides a convenient model for testing the hematopoietic powers of cell populations.

Animals irradiated with lethal doses of X rays can be successfully treated by intravenous injection of a suspension of hematopoietic cells from bone marrow or embryonic liver. Injection of spleen cells also gives a protective effect in the treatment of irradiated mice. Regeneration of the hematopoietic tissue of irradiated animals can also be obtained by injection of cells from the peritoneal exudate and peripheral blood (Micklem and Loutit, 1966).

It has been shown that this protective effect is due to repopulation of the depopulated stroma of the recipient's hematopoietic organs by the donor's cells (Barnes et al., 1959).

For this reason, the hematopoietic properties of cells from cultures can be tested by their protective effect when transplanted into an irradiated animal. More accurate quantitative results can be obtained by the use of the technique of counting hematopoietic colonies in the spleens of irradiated mice. After transplantation of a suspension of hematopoietic cells into irradiated mice, foci of hematopoiesis have been found to appear in the recipients' spleens on the 7th–10th day, each characterized by a particular type of differentiation: erythroid, myeloid, or megakaryocytic (Till and McCulloch, 1961). On the average, 10^4 cells of syngenetic bone marrow must be transplanted in order to form a single splenic focus. Each focus arises from a single stem cell; i.e., it is a clone.

Stem cells forming foci not only enable differentiated hematopoietic forms to appear, but also enable them to maintain themselves. If they are within a focus of a specific nature, they retain their polypotency, i.e., their ability to differentiate in various directions. If a suspension prepared from erythroid foci is injected into an irradiated recipient, foci of all types develop in the spleen. It is possible to judge whether a transplanted suspension contains hematopoietic stem cells from the formation of hematopoietic foci in the spleen, just as it is possible to judge from the ability of the irradiated recipients to survive.

Until recently, there was no reliable experimental evidence to show that hematopoietic cells could be kept—much less, that they could proliferate—for any length of time in culture. This situation applied to attempts to grow hematopoietic cells in both monolayer and suspension cultures.

The first attempt to treat irradiated animals by injecting fibroblasts from long-term monolayer bone marrow cultures was made by McCulloch and Parker (1957). Mice irradiated at a dose of 400 R received an intravenous injection of cells from bone marrow cultures, as a result of which 20% of the recipients survived, compared with 5% in the untreated controls. The model used was not suitable for assessing the protective power of the culture cells, however, because the dose of irradiation was too small and did not lead to 100% mortality among the control animals. The increase in the percentage of surviving recipients could be explained not only by the direct participation of the injected cells in hematopoiesis, but also by their stimulant effect on regeneration of the recipients' own hematopoietic tissue.

In another investigation (Monit and Sato, 1967), spleen cells from young mice were grown for 4 days in medium to which 15% horse serum and 2.5% fetal calf serum had been added. After 4 days, cells from the cultures were injected intravenously in a dose of 1.5×10^7 into lethally irradiated syngenetic mice, and the procedure resulted in protection of the animals. This investigation shows that stem cells can in fact be preserved *in vitro* for 4 days, and after a period in an irradiated animal, they can again be grown in culture for a further 4 days, and so on. These results, however, still give no evidence to show that stem cells can survive *in vitro* for more than 4 days.

The results obtained by Billen (1957) demonstrate that bone marrow from suspension cultures, after growth *in vitro* for several days, can protect an irradiated recipient. A suspension of mouse bone marrow cells was grown in culture by the method of Osgood and Brownlee. An intravenous injection of $0.5-2 \times 10^6$ cells from the cultures was given to syngenetic recipient mice irradiated at a dose of 900 R. Survival of a considerable proportion of the recipients was observed after treatment with cells grown in suspension cultures for 4–7 days. If the bone marrow was grown in culture longer than this, its protective potential diminished sharply.

Bone marrow explanted by Reisner's method did not give better protection to irradiated mice than bone marrow from suspension cultures. For instance, after injection of cells from 4-day cultures, 100% of recipients irradiated at a dose of 900 R survived. On the 6th day of cultivation, the bone marrow enabled only 18–44% of the irradiated mice to survive, while after a longer period of cultivation, it completely lost its protective properties (Billen, 1959).

In the course of cultivation, young forms of cells are replaced by more mature forms. In the original cell suspension, for instance, myeloblasts and myelocytes accounted for 14% of the total number of cells, falling to 10% on the 4th day and to 6% on the 9th day of cultivation, whereas the percentage of mature granulocytes increased. On the 9th day of cultivation, macrophages, fibroblasts, and unidentified cells accounted for 69%, whereas in the original material, this group accounted for only 2% of the total (Billen, 1959).

According to Dexter and Lajtha (1974), proliferation of CFUs *in vitro* is maintained for 5–9 weeks in cocultures of thymus + bone marrow cells or bone marrow + bone marrow cells in liquid media. The development of a monolayer of attaching cells appeared to be of importance for this maintenance of stem cells in culture.

Myelopoiesis is maintained for 10–14 days in agar cultures of bone marrow. In these cases, it has been shown that colonies of myeloid cells can develop because of proliferation and differentiation of committed precursors that are differentiating in the myeloid direction; if hematopoietic cells are grown on a feeder layer of embryonic kidney cells, the number of these precursors may actually increase above the initial level for 1 or 2 days, after which it begins to fall, so that by the end of the 4th day, the cultures contain virtually no precursors capable of giving rise to new colonies if explanted into fresh nutrient medium (McCulloch and Till, 1971). The question arises whether hematopoietic stem cells are preserved in such a system and, if they are, for how long and in what state? The method of explantation of hematopoietic cells in agar creates difficulties when used for retransplantation experiments *in vivo* (Metcalf, 1968). These difficulties can be overcome by use of the modified agar culture method suggested by Worton *et al.* (1969). Bone marrow cells suspended in liquid nutrient medium with 10% serum (not containing agar) are placed over a layer of 0.5% agar containing added feeder cells or conditioned medium. Under these circumstances, the bone marrow cells are in agar and suspension cultures simultaneously. Together with the layer of liquid nutrient medium, they can easily be removed from the culture vessel, washed, and injected intravenously into an irradiated recipient. The number of colony-forming cells falls sharply during the first few days after explantation, whether grown on a feeder layer or in the presence of conditioned medium. It is greater (25%) in cultures on a feeder layer on the 2nd day, however, than in cultures on conditioned medium, in which colony-forming cells number 5–10% of the initial value on the 2nd day and disappear completely by the end of the 4th day.

In both cases, the hematopoietic stem cells persist for only a short time (a few days) in the cultures. In what state do they persist? In experiments in which increasing doses of thymidine-H^3, causing death of cells in the S-period, were added to the nutrient medium, it was found that most of the hematopoietic stem cells in 2-day cultures growing on a feeder layer were in a state of active proliferation, whereas the stem cells of cultures with conditioned medium had left the proliferating pool: their number was virtually unchanged after exposure to thymidine-H^3 (McCulloch and Till, 1971). These clear observations indicate that the conditions of cultivation significantly affect the state of hematopoietic colony-forming cells. One cannot rule out the possibility that in such cases, the conditions of cultivation determine which part (proliferating or resting) of the stem cell population remains.

The work of Testa and Lajtha (1972) has shown that the number of colony-forming cells in agar cultures of bone marrow from normal mice falls over a period of 4 days to 10% of its initial level. During the first 24 hr after explantation of the bone marrow of donors previously irradiated at a dose of 450 R, there is a small increase in the number of colony-forming units (CFUs). On the following days, however, their number gradually falls, although on the 4th day, it is still higher than the number of CFUs in the cultures of bone marrow from normal mice. These results indicate that the CFUs of regenerating bone marrow evidently survive better in culture if they are in mitotic cycle than do resting CFUs of unirradiated bone marrow.

Colony-forming cells thus persist only for the first few days in agar cultures, in which foci of myeloid cells are found for a period of 10–14 days.

By the use of a method of culture in a thin layer of agar, with a monolayer of irradiated embryonic fibroblasts as a feeder layer, Dick and Van Bekkum (1972) showed that the number of polypotent colony-forming cells in 5-day mouse bone marrow cultures is greater than their number in the original material. In their experiments, these workers used bone marrow fractions from mice treated with vinblastin–mustine. This treatment stimulates the entry of stem cells into proliferation. Dick and Van Bekkum consider that colonies arise in their system *in vitro* not by proliferation of committed precursors, as takes place in ordinary agar cultures, but from polypotent colony-forming cells that also produce colonies in the spleens of irradiated mice. The presence of CFUs in colonies growing in agar cultures does not prove, however, that the myeloid colonies in fact develop from them. It is possible to imagine a situation in which the CFUs persist within the colonies, where they find adequate conditions for proliferation, while the colonies themselves nevertheless develop from committed myeloid precursors.

In organ cultures of embryonic liver, hematopoiesis is maintained for months (Luriya *et al.*, 1969a–c; Chertkov *et al.*, 1974). It will be asked whether hematopoietic stem cells are maintained under these conditions. The duration of hematopoiesis in such cultures shows indirectly that hematopoietic stem cells are present in the population, and they gradually embark on a course of differentiation, because the duration of hematopoiesis *in vitro* is much greater than the duration of proliferation and differentiation of any non-self-supporting precursors.

To determine the number of hematopoietic colony-forming cells in organ cultures (Latsinik *et al.*, 1969), the method of obtaining hematopoietic colonies in the spleens of irradiated mice was used. Cell suspensions were prepared from liver explants from mouse embryos aged 17–20 days. After cultivation for 7–23 days, the suspensions were injected intravenously in a dose of 4–6×10^4 cells into syngenetic mice irradiated at a dose of 850 R. The mice were sacrificed on the 9th–10th day of transplantation, and the number of hematopoietic colonies in their spleens was counted (Fig. 22).

Cells from cultures aged 7–23 days, after injection into irradiated mice, formed typical large hematopoietic colonies in the spleen (Table 8). In their composition and structure, these foci were typical hematopoietic foci such as are characteristic of the spleens of irradiated mice injected with hematopoietic cells. Most of the foci were erythroid. In the spleens of mice into which cells were transplanted from 13-day cultures, 77.9% of the colonies were erythroid, 10.7% granulocytic, 6.1% megakaryocytic, and 5.3% mixed (Latsinik *et al.*, 1969).

In organ cultures of the embryonic liver, not only do the hematopoietic cells proliferate and mature for at least 3 weeks, but also a line of colony-forming cells, evidently hematopoietic stem cells, is maintained.

Judging from the number of colonies, the number of CFUs in explants of embryonic liver is relatively high (Table 8).

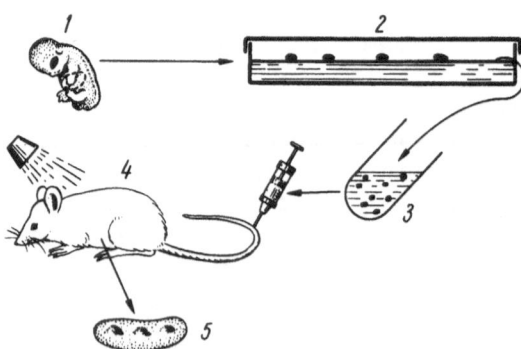

Fig. 22. Scheme of injection of cells grown in organ cultures of embryonic liver into irradiated recipient mice: (1) livers of 17–20 day mouse embryos; (2) growth in organ cultures for 7–23 days; (3) preparation of a cell suspension from cultures; (4) intravenous transplantation of cells into recipient mice irradiated at a dose of 850 R; (5) colonies of hematopoietic tissue in spleens of irradiated mice on 10th day after transplantation.

TABLE 8. Colony-Forming Units in Cultures of Embryonic Liver[a]

Age of embryos (days)	Number of cultures	Duration of cultivation (days)	Number of viable nucleated cells added ($\times 10^3$)	Number of colonies per spleen	Number of CFUs per 10^5 cells
17	25	7	66	8, 13, 16, 20, 21	23.6
17	25	7	132	25, 28, 31, 32, 32	22.3
19	48	12	54	17, 20, 21, 21, 22, 22, 23	21.7
17	70	13	38	0, 10, 14, 18, 19, 20, 22, 25, 25, 25	44.2
17	65	23	16.5	0, 0, 2, 3, 5	12.0
—	—	—	0	0, 0, 0, 0, 0, 0, 0, 0, 0, 0, 0, 0, 0, 0, 0, 0, 0, 0, 0, 0, 0, 0, 0, 0, 0, 0, 1	—

[a]After Latsinik *et al.* (1969).

The concentration of CFUs in organ cultures is higher than in the original material. It continues to rise until the 13th day of cultivation, when it is 44.2 per 10^5 cells. In 23-day cultures, the number of CFUs falls to 12.0 per 10^5 cells, i.e., close to the characteristic level for the original material (embryonic liver).

The concentration of colony-forming cells increases because hematopoietic stem cells are not only used up, but also proliferate in organ cultures. The intensity of proliferation of the stem cells in organ cultures is also indicated by figures for the effect of vinblastin on the number of CFUs in organ cultures of embryonic liver (Figs. 23 and 24). During exposure of cultures aged 7–17 days to vinblastin for 24 hr, the number of CFUs is reduced to 90% (Luriya *et al.*, 1971c; Table 9).

These results show that about 90% of the hematopoietic stem cells in organ cultures aged 7–17 days take part in cell division during a 24-hr period.

More detailed observations on the kinetics of hematopoietic stem cells (CFUs) in organ cultures have recently been reported (Chertkov *et al.*, 1974). In the first 2 weeks, the number of hematopoietic stem cells in explants of mouse embryonic liver falls to approximately one-tenth of the original level. A steady state is then established, and for a month, the number of CFUs remains at this level. Throughout this period, a definite ratio is maintained between the number of stem cells and of their progeny, and, in particular, the number of mature

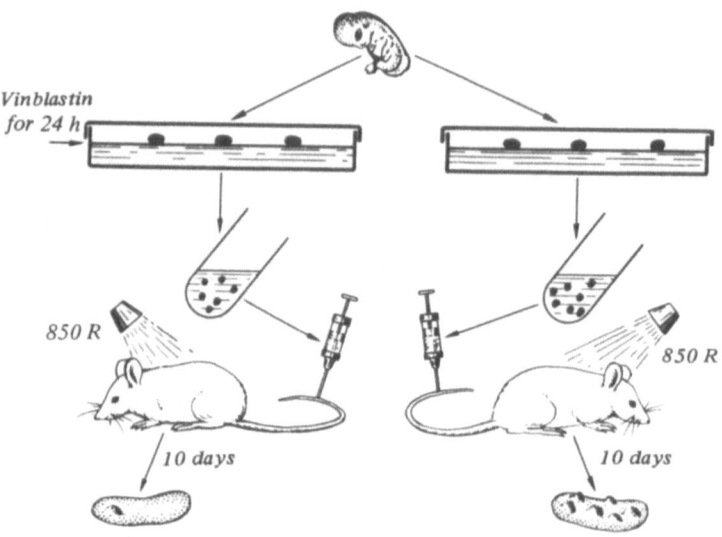

Fig. 23. Action of vinblastin on colony-forming cells of an embryonic liver culture.

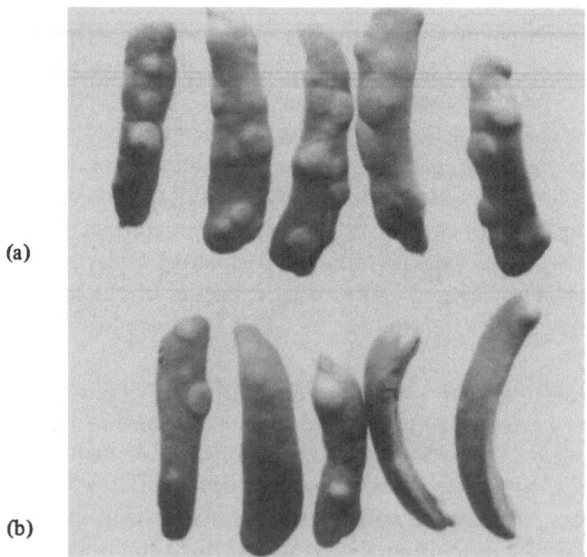

Fig. 24. Hematopoietic colonies in spleens of irradiated recipient mice receiving transplanted cells from 14-day organ cultures of embryonic liver: (a) spleens of mice after transplantation of cells from cultures on ordinary medium; (b) spleens of mice after transplantation of cells from cultures grown in medium with added vinblastin.

TABLE 9. Action of Vinblastin on Colony-Forming Units in Cultures of Embryonic Liver[a]

Duration of cultivation (days)	Group of animals	Number of mice	Number of cells injected per mouse	Number of colonies per spleen	Number of CFUs per 10^3 cells injected	Percentage of cells dying under the influence of vinblastin
1	Control	9	35,000	1.9±0.8	5.4	70
	Experimental	10	102,000	1.6±0.5	1.6	
7	Control	8	51,000	23.4±1.7	45.5	87
	Experimental	5	70,000	4.2±0.5	6.0	
12	Control	7	32,000	14.4±1.0	45.0	56
	Experimental	8	126,000	25.0±3.3	20.0	
17	Control	6	43,000	9.7±1.6	22.6	
	Experimental	7	78,000	2.0±0.8	2.6	

[a] After Luriya et al. (1971).

granulocytes with a limited life span in culture. As has been pointed out already, the relative content of CFUs at this time is higher than in the original liver (about $40-50/10^5$ cells compared with $6-10/10^6$ in the liver of 18-day embryo). After the 40th day of cultivation, the number of stem cells starts to fall again, but like the more mature hematopoietic cells, a definite number of them remains for a further month.

It has been confirmed by the thymidine suicide method that during cultivation, the hematopoietic stem cells continue to proliferate actively: the degree of suicide is 30–45% after exposure of the cells for 20 min to thymidine-H^3 with high specific activity. Only after 1.5 months has elapsed, during the final decrease in the number of CFUs, does the rate of their proliferation slow down and the degree of thymidine suicide fall to 15–25%.

Consequently, conditions are created in organ cultures of the embryonic liver that simulate stable hematopoiesis *in vivo* to some degree. Proliferation and differentiation of the stem cells are maintained, so that the total number of stem cells remains constant. The mechanism of this stabilization, however, is evidently different from that *in vivo*.

In the adult organism, stem cells (CFUs) proliferate slowly in the steady state, whereas their committed progeny (CFUc's) participate in the cell cycle. In organ cultures, on the other hand, the steady state is maintained by rapid proliferation of the stem cells themselves.

Hematopoietic stem cells survive much less successfully in organ cultures of adult mouse bone marrow. Their number falls by 10^4 times in the first 10 days of cultivation, and at no time can proliferation of the stem cells be found (by the thymidine suicide method). Explantation of regenerating mouse bone marrow into organ cultures gives somewhat better results if the regeneration is induced by irradiation at a dose of 250 R. In this case, hematopoietic stem cells continue to proliferate after explantation, and hematopoiesis persists a little longer. If, however, bone marrow is explanted from early radiation chimeras (6 days after lethal irradiation and transplantation of syngenetic bone marrow), hematopoiesis comes to an end even sooner in the organ culture than after explantation of the bone marrow of normal mice (Chertkov *et al.*, 1974).

It can accordingly be concluded that the inferior growth of adult bone marrow is not entirely due to the fact that its stem cells, unlike those of embryonic liver, do not proliferate. The properties of the stroma bordering on the hematopoietic cells evidently play an important role. In the system described above, when bone marrow from normal mice was explanted onto previously grown embryonic osteogenic tissue (see p. 53), the hematopoietic stem cells began to proliferate (the degree of thymidine suicide was 15–25%), and their number at all times of cultivation was approximately one-tenth of that in cultures without an osteogenic feeder layer (Chertkov *et al.*, 1974).

Erythropoiesis in embryonic liver *in vitro* is known to differ from erythropoiesis in bone marrow in certain morphological features, and also in its sensitivity to erythropoietin (Lucarelli *et al.*, 1968; Jacobson *et al.*, 1959). Erythropoiesis in the adult organism is erythropoietin-dependent; i.e., the presence of erythropoietin is essential for its maintenance. Erythropoiesis in embryonic liver is less sensitive to the effect of polycythemia than medullary hematopoiesis. Transplanted bone marrow cells do not lead to the development of erythroid colonies, for instance, in plethoric recipients (Feldman and Bleiberg, 1967).

In case of injection of embryonic liver cells into irradiated polycythemic recipients, colonies predominantly of the erythroid type are formed in the spleen.

Do the stem cells of embryonic liver, which proliferate under the conditions of organ culture, preserve their original insensitivity to polycythemia? With the use of a model of transplantation into irradiated polycythemic recipients, it was shown that in organ culture, the colony-forming cells of embryonic liver become sensitive to the inhibitory effect of polycythemia, i.e., that they acquire the characteristic property of the stem cells of the adult organism (Latsinik *et al.*, 1970*b*). Cell suspensions obtained from 13-day organ cultures of mouse embryonic liver were injected into normal and polycythemic irradiated mice. In the polycythemic animals, there was a marked decrease (by approximately two-thirds) in the total number of colonies in the spleen, and erythroid foci were absent.

Prolonged maintenance of a line of hematopoietic stem cells and their intensive proliferation were thus obtained for the first time in organ cultures. These results show that hematopoietic stem cells can maintain themselves without influences reaching the hematopoietic organs from the rest of the body (this problem remained unsolved until very recently). At the same time, they show that organ cultures provide a promising method of cultivating hematopoietic tissue, including hematopoietic colony-forming cells.

Organ cultures have an important limitation as a culture method: difficulties arise as a result of the production of a large cell mass. Modern organ culture techniques thus cannot be regarded as suitable for the cultivation of hematopoietic tissue cells for further transplantation.

The situation with regard to monolayer cultures of hematopoietic tissue is different. Fibroblast-like cells growing in these cultures proliferate actively and are readily passaged. Accordingly, the problem of obtaining a large cell mass in these cultures encounters no difficulty in principle; the problem is organizational rather than scientific in nature. Can fibroblast-like cells from monolayer cultures be used for transplantation? In other words, what are the possibilities for differentiation of these cells on retransplantation *in vivo*?

Diploid strains of fibroblast-like cells have now been successfully obtained from the bone marrow, spleen, thymus, and lymph nodes of guinea pigs, rabbits, and man (Fridenshtein *et al.*, 1970*a–c*; Miskarova *et al.*, 1970; Panasyuk *et al.*, 1972). All consist of fibroblast-like cells, similar in their morphology. The question is: what is the histogenetic potential of the cells of these strains? To answer it, various methods of transplantation of the cells *in vivo* can be used: transplantation in diffusion chambers, intravenous injection into irradiated animals, and transplantation of aggregated cells under the kidney capsule.

Fridenshtein *et al.* (1970*a,b*) have studied the ability of fibroblasts from monolayer cultures of the spleen and bone marrow to differentiate by transplanting cells grown in cultures of the spleen and bone marrow of guinea pigs in diffusion chambers. They introduced 2×10^7 fibroblast-like cells from bone marrow culture into a diffusion chamber, and after 20–30 days observed intensive osteogenesis in the chamber: compact bone with immured osteocytes, deposition of ground substance, and a surface layer of osteoblasts were formed (Fig. 25). If fewer cells (2×10^6) were placed in the diffusion chamber, no osteogenesis was observed. The fibroblast-like cells grown in monolayer cultures of bone marrow are thus not the final form of irreversible differentiation. These cells preserve the powers of differentiation of marrow stroma, which, if transplanted subcutaneously, under the kidney capsule, or in a diffusion chamber, gives rise to osteogenic tissue (Fig. 26; Fridenshtein *et al.*, 1968).

Fibroblast-like spleen cells grown in monolayer cultures, if transplanted into a diffusion chamber, form a network of connective tissue cells; osteogenic differentiation is not observed under these conditions.

Osteogenesis observed in diffusion chambers after transplantation of fibroblast-like cells from bone marrow cultures is spontaneous; i.e., it arises through interaction among the cells of the same population and does not require any induction by external agents.

The potential osteogenic properties of the bone marrow fibroblasts do not disappear on prolonged subculture. For instance, following transplantation in

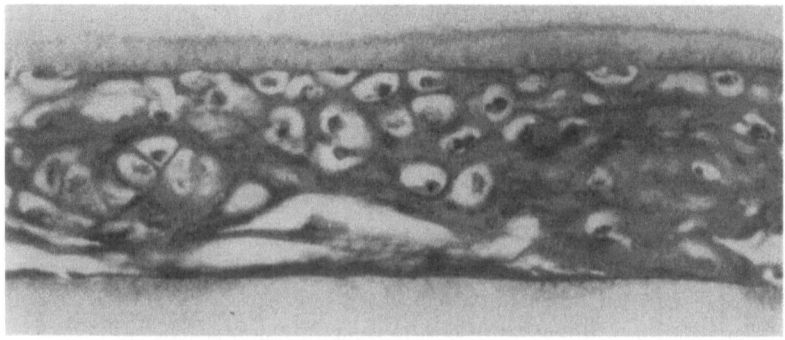

Fig. 25. Osteogenesis in a diffusion chamber with fibroblasts from a bone marrow culture. Section. After Fridenshtein *et al.* (1970*b*).

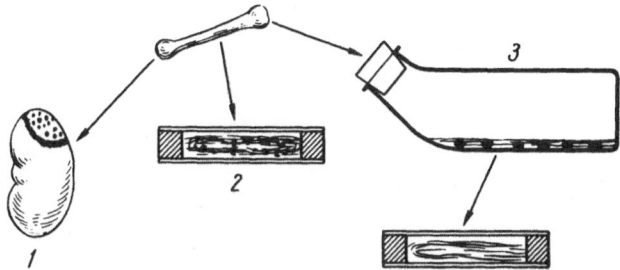

Fig. 26. Experimental methods for the analysis of the stroma of hematopoietic (and lymphoid) organs: (1) transplantation under the kidney capsule; (2) transplantation in a diffusion chamber; (3) explantation in monolayer cultures followed by transplantation in a diffusion chamber.

diffusion chambers of rabbit bone marrow fibroblasts, after 17 passages *in vitro*, the formation of mature bone tissue was observed in the chamber on the 28th day (Miskarova *et al.*, 1970).

Osteogenesis can be induced by transitional epithelium of the urinary bladder (Huggins, 1931; Fridenshtein, 1963). It has been shown by the use of an isolated system (diffusion chamber) for the transplanted cells that cells from the spleen, peripheral blood, and peritoneal exudate are populations in which osteogenesis can be induced (Lalykina and Fridenshtein, 1969). It was important to determine whether osteogenesis can be induced in fibroblast-like cells grown in cultures of bone marrow and spleen. Experiments showed that osteogenesis can be successfully induced in diffusion chambers by growing transitional epithelium of the urinary bladder together with fibroblasts from monolayer cultures of bone marrow and spleen. The concentration of fibroblasts from bone marrow cultures, which themselves cannot undergo spontaneous osteogenesis, gives rise to osteogenesis under the influence of the transitional epithelium. Intensive osteogenesis also takes place during combined cultivation of fibroblasts from spleen cultures with urinary bladder epithelium (Fig. 27). The population of fibroblast-like cells obtained in monolayer cultures thus possesses considerable differentiation potential, which can be manifested either spontaneously or through the influence of an inducer (Fridenshtein *et al.*, 1970*b*).

Transplantation of fibroblasts from cultures of rabbit bone marrow and spleen under the kidney capsule has shown that these cells can transfer the specific microenvironment characteristic of a source hematopoietic organ. In actual experiments (Fridenshtein *et al.*, 1974*a*) in which marrow fibroblasts were passaged 4–10 times *in vitro*, removed with trypsin, and transplanted as a cell clump under the kidney capsule (of the same rabbit from which the original explanted bone marrow had been obtained), 1 month later, bone trabeculae with proliferating myeloid cells among them could be seen under the kidney

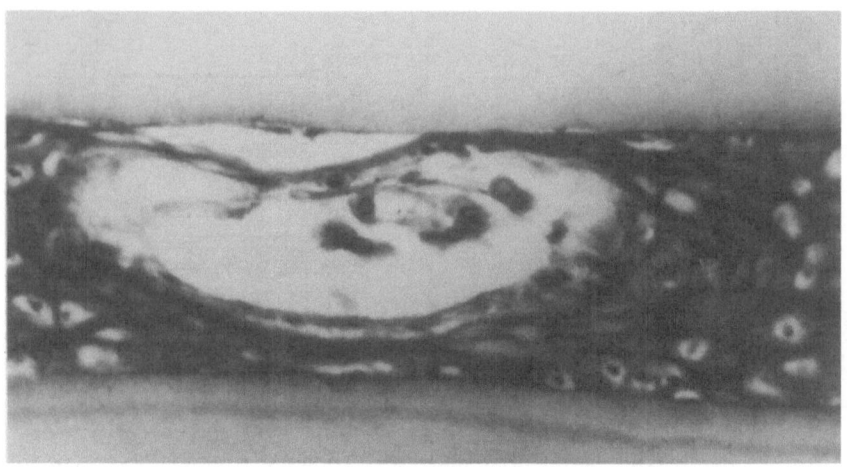

Fig. 27. Osteogenesis in a diffusion chamber from fibroblasts from a culture of spleen cells together with transitional epithelium. After Fridenshtein *et al.* (1970*b*).

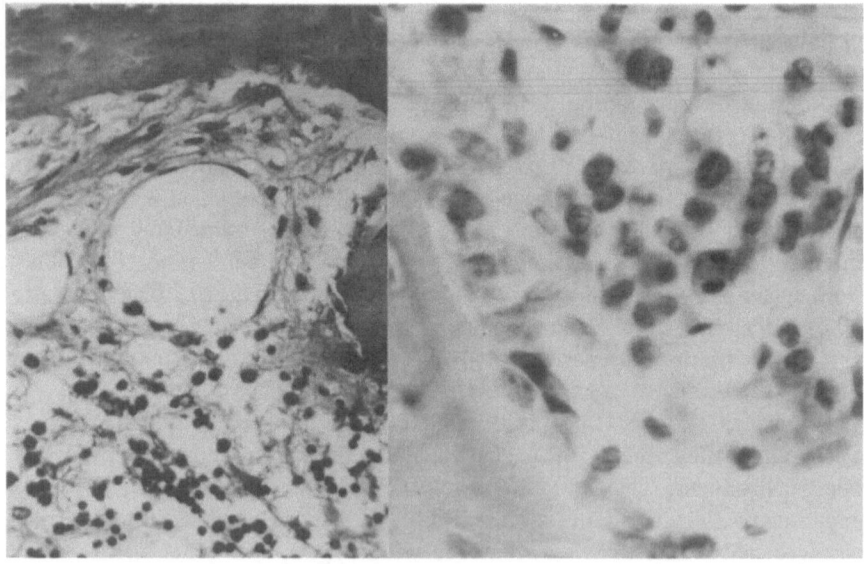

Fig. 28. Formation of osseous and myeloid tissue beneath the capsule of the rabbit kidney at the site of transplantation of fibroblasts from a culture of bone marrow cells after seven previous passages. After Fridenshtein *et al.* (1974*a*).

capsule (Fig. 28). Similar transplantation of fibroblasts obtained from spleen cultures led to the formation of a lymphoid organ containing lymphoid follicles under the kidney capsule. After heterotopic transplantation of fragments of bone marrow or lymphoid organs (Fridenshtein *et al.*, 1968; Didukh and Fridenshtein, 1970; Fridenshtein and Kuralesova, 1971), restoration of hematopoiesis or lymphopoiesis in the grafts depended on preservation of the donor's stromal cells in the established grafts, thereby creating a microenvironment favorable to repopulation by the recipients' hematopoietic and lymphoid cells. The latter are promoting their differentiation in the characteristic direction of the transplanted organ.

The results of transplantation of fibroblasts from monolayer cultures of bone marrow and spleen given above indicate that lines of stromal cells, responsible for transferring the specific microenvironment, are maintained in these cultures.

It can be concluded that there is now a real possibility of growing stromal cells of hematopoietic and lymphoid tissue. The problem of their possible use in transplantation will evidently be solved in the not too distant future.

CHAPTER III

Lymphoid Tissue *In Vitro*

1. Lymphoid Tissue Explanted in a Plasma Clot

The method of explantation in a plasma clot has yielded valuable data on the behavior of lymphoid tissue and its constituent elements *in vitro*.

Spleen

The first investigation into explantation of lymphoid tissue was carried out in 1910 when Carrel and Burrows (1910*b*) grew fragments of a kitten's spleen in cat plasma. They found that after 24 hr erythrocytes, granulocytes, and lympho-cytes migrate from the explant. On the 2nd and 3rd days, large cells capable of phagocytosis begin to appear in the cultures, while the lymphocytes gradually disappear. Later, the zone of growth becomes connective tissue in character: the cells in it are arranged as rays stretching from the center to the periphery of the plasma clot.

Similar results on the whole were obtained by other workers who clarified some details of the behavior of the cells in the zone of growth. Cells with a morphology intermediate between lymphocytes, on the one hand, and reticulum cells and macrophages, on the other hand, were found in cultures of rabbit spleen (Maksimov, 1916). These pictures, together with the proliferation of fibroblasts in older cultures, provided the basis for the hypothesis that in tissue culture, lymphocytes are transformed into polyblasts, which in turn are trans-formed into fibroblasts (Maksimov, 1916). Movement of the lymphocytes in the zone of growth is ameboid in character (Fazzari, 1926) and does not last long: in cultures of spleen cells from chick embryos and chickens, for example, it lasts only for the 1st day. The composition of the zone of growth on the first days after explantation depends on the age of the donor animal: in cultures from

14-day chick embryos, hematopoietic cells migrate into the zone of growth, while mainly lymphocytes migrate from the spleen of adult hens. In the later stages, histiocytes and fibroblasts appear in the same cultures, and the migrating lymphocytes and hematopoietic cells die (Fischer, 1930).

After three or four subcultures, the number of free cells in the cultures decreases sharply, while after six subcultures, the original tissue becomes spongy and the fragment becomes round and is surrounded by a zone of connective tissue growth. After eight subcultures, the culture acquires the typical appearance of pure cultures of connective tissue (Fischer and Doljanski, 1929).

Cases of spontaneous formation of small foci consisting of 6–8 polymorphic cells (round, rectangular, polygonal) of different sizes have been described during prolonged cultivation of spleen cells. Sometimes these cells are numerous. They can invade the cytoplasm of the fibroblasts, pass through it, and come out on the other side (Fischer and Doljanski, 1929). The connection between these cells and lymphocytes has not been explained.

Lymph Nodes

In his classic investigations, Maksimov cultivated fragments of lymph nodes in a plasma clot. Maksimov studied his cultures by means of intravital staining techniques and histological investigation. He concentrated his attention on cell transformations. Although he could not employ modern methods of experimental analysis (chromosome markers, thymidine labeling, isospecific sera, etc.), which did not become available until several decades later, he nevertheless was able to draw some important conclusions, many of which have subsequently been confirmed by these modern methods, while others are still stimulating experiments to prove or disprove them. Although Maksimov's work was done in the 1920s (Maksimov, 1922, 1923*a,b*), it is still as up to date as ever.

During the 1st day, explanted fragments of lymph nodes are surrounded by a zone of cells performing ameboid movements at a speed of 0.03 mm/min and forming pseudopodia (in cultures of human lymph nodes) (W. H. Lewis and Webster, 1921). Most of the emigrating cells are small lymphocytes, which degenerate on the following days. Some lymphocytes near the explant are transformed, in Maksimov's opinion, into plasma cells, and many reticulum cells are liberated from the syncytium and converted into fixed or free macrophages. These cells are distinguished by a characteristic vesicular nucleus, and their cytoplasm often contains a pigment that stains green with azure-eosin.

Later, many cells morphologically identical with the fibroblasts of adult connective tissue appear in the cultures. Maksimov considered that fibroblasts arise from cells of the reticular syncytium lying at the periphery of the explants. In the course of cultivation, hypertrophied lymphoid cells are formed inside the fragment and in the regions adjacent to its border. Morphologically, there are intermediate cells for two types of transformation: (1) small lymphocytes into large (of the lymphoblast type); and (2) monocytes into macrophage–polyblasts, i.e., cells similar to those that appear *in vivo* during inflammation. Over the course of time, the number of lymphocytes decreases, and the explant changes into a culture containing a small but varied number of fibroblasts.

That was the picture of cultivation in plasma. Addition of bone marrow extract to the plasma lengthens the life of the lymphocytes in cultures of rabbit lymph nodes. In the strongly basophilic proliferating lymphocytes grown under these conditions, a pigment usually characteristic of reticulum cells appears. This observation was responsible for the hypothesis that lymphocytes can arise *in vitro* from the reticular cells of the stroma. Further experiments are of course necessary to prove or disprove this possibility, which now seems very unlikely. The same conclusion applies to another observation by Maksimov.

In some cultures of rabbit lymph nodes in plasma and bone marrow extract, myeloid cells and megakaryocytes were seen to be formed. They appeared after 4 days in those cultures, which had intensive growth of fibroblasts at their periphery. The myeloid cells were found not in the zone of migration, but in the explant itself. Maksimov postulated that the myelocytes were derived from large basophilic lymphocytes. The lymphocytes and hematopoietic cells had only a short life in the cultures. Over the course of time, only connective tissue cells and giant multinuclear cells remained in the explants.

It must be emphasized that although the only known attempt to reproduce these results proved unsuccessful (Shiomi, 1925), the possibility that myelopoiesis may take place in cultures of lymph nodes deserves the closest attention, especially because it is difficult to reconcile with modern views regarding the localization and morphology of myeloid precursor cells.

Thymus

The chief morphological feature that distinguishes the thymus from other organs of the lymphoid system is the presence in its stroma of an epithelial component, which is clearly apparent during culture in a plasma clot. For this reason, analysis of the epithelial zone of growth of the explants provided an important basis for proof of the ectodermal nature of the epithelium of the thymus (Khlopin, 1946).

Cultures of the thymuses of amphibians, reptiles, birds, and mammals behaved on the whole similarly in plasma cultures (Chasovnikov, 1927; Popov, 1927; Galustyan, 1940; Murray, 1947). Proliferation of the epithelium and migration of "polyblasts" and phagocytes are the primary features. Migration of lymphocytes from the thymus takes place gradually. Many of the migrating lymphocytes die, others, perhaps, are converted into lymphoblasts, and some form

compact groups. After emigration of the lymphocytes, the epithelial components of the fragment are more clearly distinguishable. The explant is surrounded by wide epithelial membranes, sheets and bands of epithelium, and a
zone of connective tissue growth. The number of lymphocytes in the cultures
later decreases sharply, after which they disappear completely.

On the 3rd or 4th day of cultivation, epithelial cells are grouped into
compact foci. Cysts formed as a result of the secretory activity of the epithelium
appear among the cells. Structures resembling Hassall's corpuscles can be found
among the cells of the epithelial syncytium and in the zone of liquefaction of
the plasma. The ratio between the numbers of fibroblast-like and epithelial cells
in the zone of growth may vary considerably. Despite growth of the epithelium,
liquefaction of the fibrin does not take place in most cultures.

On the whole, cultivation of fragments of lymphoid organs in a plasma clot
does not lead to the long preservation or proliferation of the lymphocytes. With
the roller tube method, cultures of lymphoid tissue with a somewhat longer life
could be obtained, but it had no substantial advantages over the classic method.

During the first days of growth, lymphocytes also migrate from the fragment, and the stroma proliferates intensively to produce projections into the
surrounding medium. When fragments of the lymph node or spleen are explanted, the zone of growth consists of connective tissue, but not of lymphocytes. Instead of lymphocytes, the fragments are surrounded by macrophages
and by fibroblast-like cells that form structures resembling grass. If fragments of
thymus are explanted, on the other hand, in addition to connective tissue, the
ectodermal epithelium also proliferates actively to form membranes and to
envelop an explanted fragment. Lymphoid cells do not regenerate in these cultures.

Cells of the Peripheral Blood and Lymph

Although the peripheral blood and lymph are not directly related to lymphoid
tissue, it serves a useful purpose to examine the results of their explantation
in a plasma clot in this section. When cultivated in a plasma clot, the peripheral
blood and lymph behave similarly to lymphoid tissue, but they present a more
homogeneous cell system.

The appearance of fibroblasts and histiocytes, i.e., of cells absent in the
original material, in plasma cultures of blood cells was observed by the very first
investigators who cultivated blood *in vitro*. It could easily be seen in buffy coat
cultures from rabbits, dogs, and guinea pigs (Avrorov and Timofeevskii, 1915)
that connective tissue cells replace the explanted leukocytes; agranular leukocytes (monocytes or lymphocytes) were believed to be transformed into polyblasts and fibroblast-like cells.

Carrel and Ebeling (1922) actually obtained pure cultures of mononuclear cells from chicken blood. In some cases, they also found growth of fibroblasts. Similar results were obtained in cultures of rabbit and human blood; macrophages, fibroblasts, and epithelioid and giant cells were formed (Timofeevskii and Benevolenskaya, 1926, 1927; M. R. Lewis, 1925; Vasiliu and Stoica, 1929; Rhoads and Parker, 1928). These observations raise an important question: which blood cells—monocytes or lymphocytes—are the precursors for macrophages and for fibroblasts, i.e., can be transformed into them? This question was specially studied by Maksimov in a series of investigations carried out in the 1920s. The object of his work on buffy coat culture in a plasma clot was to study transformation of chicken, guinea pig, rabbit, and monkey blood cells *in vitro*. Transformation was studied in greatest detail in the cultures of guinea pig blood (Maksimov, 1928).

In the first few hours after explantation, cells migrate from the fragment of buffy coat, which, after 3–4 hr, is surrounded by a zone of granular and agranular leukocytes. Granulocytes move before the agranular forms. After cultivation for 10 hr, the granulocytes begin to degenerate; many cells in the fragment die at this time, while others hypertrophy and exhibit phagocytic activity. At the same time, large amoeboid cells, accumulating neutral red in the cytoplasm, appear. Several forms intermediate between these cells and small lymphocytes can be found. Maksimov attempted to judge from the density of the nucleus in these large phagocytic cells whether they originated from lymphocytes or monocytes.

Most of the granulocytes degenerate after 24 hr. Very few unchanged lymphocytes and monocytes remain in the culture; the number of large cells with amoeboid movement capable of phagocytosis increases sharply.

The predominant type of cell in the 48-hr cultures is macrophages (polyblasts). Mitoses are rare in these cells. On the 3rd–5th day, there appear individual fibroblast-like cells that, under the influence of various conditions (powerful illumination, heat) become round and turn into amoeboid cells. A population of large fibroblast-like cells, with large oval nuclei and a broad band of cytoplasm in which tonofibrils can be seen, develops on the 7th–14th day. The fibroblast-like cells proliferate actively and show numerous mitoses. As a result, foci and sheets of cells similar in their morphology to cultures of connective tissue obtained from subcutaneous cellular tissue are formed (Maksimov, 1916). On the 25th day, Maksimov (1928, 1929) observed the formation of collagen and reticulin fibers in peripheral blood cultures.

Thus, in a system consisting originally of lymphocytes and monocytes of the peripheral blood, macrophages appear first, followed by fibroblasts. Maksimov considered that he had found the gradual transformation of the lymphocyte through a polyblast into a fibroblast-like cell; he illustrates his observations with beautiful drawings. These morphological pictures are not sufficient, however, for the analysis of a process of this nature, especially as regards the origin of the fibroblasts.

There is no doubt that the precursors of the fibroblasts are among the mononuclear cells of the explanted cell population. However, the concentration of the precursor cells is evidently very low. It has recently been shown that the number of precursors of fibroblasts among the peripheral blood cells is about 1 per 10^5 cells (see p. 87). It is therefore most improbable that the conversion of these cells could be observed at all frequently without resorting to time-lapse cinematography. Maksimov's observation that growth of the fibroblast population takes place chiefly, not through transformation of the mass of the polyblast into fibroblasts, but through intensive mitotic division of those few fibroblasts that appear first in the cultures, remains very important.

Transformation of blood cells into macrophages, followed by the appearance of fibroblasts, in plasma cultures of peripheral blood has frequently been described since Maksimov's time.

The main question that arose was whether lymphocytes or monocytes are the original elements for these transformations. The answer, however, has not yet been found. As Trowell (1965) points out, the authors of 41 investigations using the plasma clot explantation technique have reached several different conclusions. In 19 cases, it was concluded that lymphocytes are the chief source of macrophages in culture; in 10 cases, it was concluded that transformation takes place from monocytes; in 8 cases, the conclusion was that both lymphocytes and monocytes take part in this process; the authors of 4 investigations reached no definite conclusion whatsoever.

In some of these investigations, moreover, suspicions were aroused that fibroblasts may have been formed in the blood cultures through "contamination" of the material with connective tissue cells from the blood vessel walls. It would seem that such cells would always be present in an explanted suspension, whatever method the experimenter used to obtain the blood: by cardiac puncture or from the peripheral blood vessels. It follows from this observation that there must always be uncertainty about the presence of precursors for fibroblasts among peripheral blood cells (Jacoby, 1965; Rangan, 1967).

On the whole, the question of the existence of precursors of connective tissue among the blood cells has proved to be more difficult than appeared initially to be the case. Nevertheless, the solution of this question is of fundamental importance to a number of important problems in hematology and immunology and to the analysis of the population structures of the tissues of the internal milieu. Explantation of peripheral blood leukocytes into monolayer cultures has yielded valuable information in this connection (see below).

The peripheral blood is not the only population of free cells that gives growth of histiocytes and fibroblasts *in vitro*. Maksimov (1917) mentions that he obtained colonies of fibroblasts from cells of a peritoneal exudate. These observations were later confirmed by Vereshchinskii (1924).

The most homogeneous system used to study transformation of lymphocytes in plasma cultures consists of lymph cells from the thoracic duct (W. Bloom, 1927, 1928). Bloom cultivated lymphocytes obtained from the thoracic duct of a rabbit in a plasma clot. The principal cells in the original explant were large and small lymphocytes, with a very few erythrocytes, eosinophilic leukocytes, and sometimes monocytes as contaminants (the last numbered less than 1%). After a few hours, the lymphocytes were replaced by larger cells with an increased content of neutral red and with lipid inclusions. By the end of the 1st day, many of the unchanged lymphocytes started to degenerate. After a week, small colonies of fibroblasts containing a very few macrophages were formed in these cultures.

Cultures of lymph obtained from the efferent lymphatics of the mesenteric lymph nodes of rabbits behaved in the same way. In this case, the original material contains virtually nothing but small lymphocytes. In this case, the impression was also obtained that lymphocytes were transformed into macrophages, and then into long cells with the morphology of fibroblasts (Bergel, 1930). These results, obtained by Maksimov's pupils, correspond to his general idea of lymphocytes as original cells for the development of histiocytes and fibroblasts in culture. This question is particularly pertinent at present in connection with the problem of hematopoietic stem cells (see below).

2. Monolayer Cultures of Lymphoid Tissue and Peripheral Blood Cells

Suspensions of cells from lymphoid organs explanted into monolayer cultures cannot give rise to long-term cultures of lymphocytes. Most cells that adhere to the surface of the glass are transformed into macrophages and histiocytes, which are later replaced by a population of fibroblast-like cells (Fig. 29). Changes in the cell composition in monolayer cultures of lymphoid cells can be studied in cultures of guinea pig spleen cells (Chailakhyan, 1970). The successive phases of the changes taking place in such cultures are in good agreement with those observed by Maksimov in plasma cultures. On the first days of cultivation, a population of histiocyte—macrophages containing very small numbers of monocytes and polymorphs in various stages of degeneration is found. On the 6th day, foci of fibroblast-like cells are beginning to appear. The foci are comprised of large cells with large nuclei, one or several nucleoli, and extensive cytoplasm. In their morphology, these fibroblasts resemble cells of human diploid strains. On the 6th day, the diameter of the foci is 0.8–1.2 mm, and the mean number of cells in each is 200; on the 9th day, the diameter is 1–1.8 mm and the mean number of cells 600; by the 12th day, the foci have increased in diameter to 1.5–2.5 mm, and the mean number of cells is 1500.

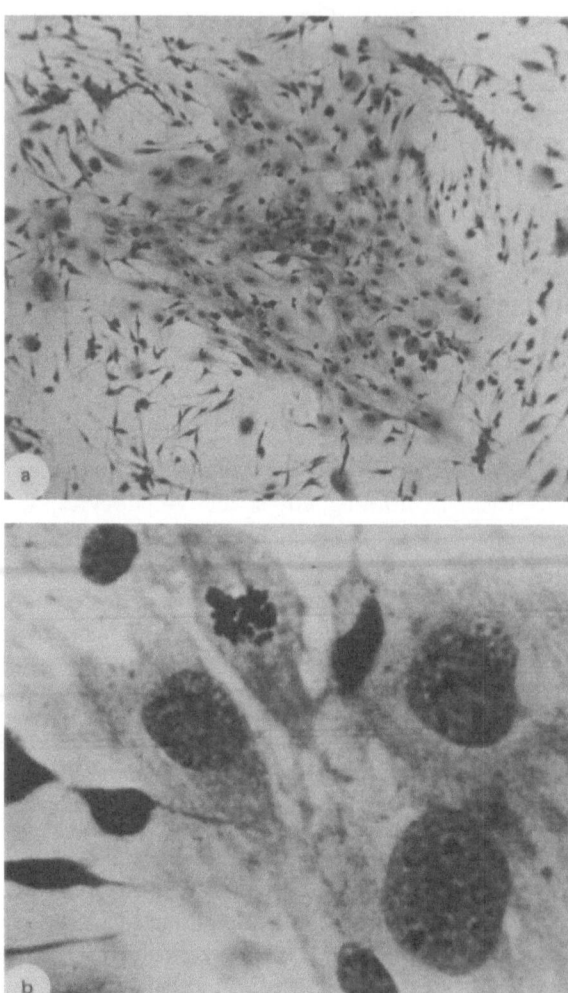

Fig. 29. Colonies of fibroblasts in monolayer cultures of mouse spleen cells. After Kuz'menko and Fridenshtein (1971).

After the 20th day, fibroblast-like cells have completely supplanted the histiocyte–macrophage population.

Each focus of this type is a cell clone; i.e., it originates from a single precursor cell (Fridenshtein *et al.*, 1970*b*). That it does is shown by results obtained by chromosomal analysis performed on foci arising in monolayer cultures from two donors differing in their sex chromosomes. A further indication of the clonal nature of the foci is the linear relationship between the number of foci and the number of explanted cells, which is valid if the number of explanted cells does not exceed 10^7 per 100-ml flask. The size of the CFU, i.e., the number of explanted cells per colony, was determined. For cultures of guinea pig spleen, the CFU is 2×10^5. This means that among 2×10^5 cells, there is one precursor that gives rise to a colony of fibroblasts if grown in monolayer culture.

Formation of colonies of fibroblasts is also observed after explantation of cell suspensions from other lymphoid organs of the guinea pig, e.g., the thymus and lymph nodes, and also during cultivation of peritoneal exudate cells (Luriya *et al.*, 1972a) and mouse spleen cells (Fig. 29). Although the phase of macrophagal growth ceases almost completely after explantation of suspensions from the thymus and lymph nodes, the dynamics of formation and growth of the colonies is the same in all cases.

The cell populations mentioned above thus contain precursors of colonies of fibroblasts. Their concentration (to judge from cloning results *in vitro*) in guinea pigs is as follows (per 10^5 cells): 9 for peritoneal exudate cells, 0.5 for spleen cells, 0.2 for lymph node cells, 0.05 for thymus cells—but 30 for cells from the pleural cavity (Fridenshtein *et al.*, 1973b). FCFC concentration may vary markedly. For example, their concentration in peritoneal exudate 20 days after subcutaneous injection of Freund's adjuvant is increased by 70 times (Luriya *et al.*, 1972a). Cells forming colonies of fibroblasts in cultures of spleen and lymph nodes belong to the category of stromal precursor cells. Cloning *in vitro* has revealed this new category of cells and has enabled the changes in their concentration in the lymphoid organs under different conditions to be estimated. For example, the number of FCFCs in a regional lymph node of guinea pigs 24 hr, 2 days, and 7 days after subcutaneous injection of diphtheria toxoid was 30 times normal; 12 days after the injection, it was 10 times normal; even 28 days after the injection, it was still approximately 5 times higher than normal (Fridenshtein *et al.*, 1974a).

By the 10th–14th day after explantation of peripheral blood leukocytes of a guinea pig in monolayer cultures, colonies of fibroblast-like cells can also be observed to be formed (Fig. 30) (Luriya *et al.*, 1971a). During the 1st day of growth, all the blood cells can be seen on the slide, but most subsequently degenerate. The few cells that remain have the morphology of histiocytes. On the 2nd day, the first colonies of fibroblasts begin to appear, and they consist of a few cells. On the 4th day, these colonies are more numerous and are larger in size. On the 10th–12th day, large colonies of fibroblasts containing up to several thousand cells can be seen with the unaided eye (Figs. 30 and 31). The colonies consist of typical fibroblasts, forming collagen fibers and containing tonofibrils in the cytoplasm. A definite linear relationship can be seen between the number of foci and the number of explanted cells. The colony-forming unit for cultures of blood leukocytes from guinea pigs varies from 2.5×10^4 to 3×10^5. It is important to note that the number of cardiac punctures involved in taking the

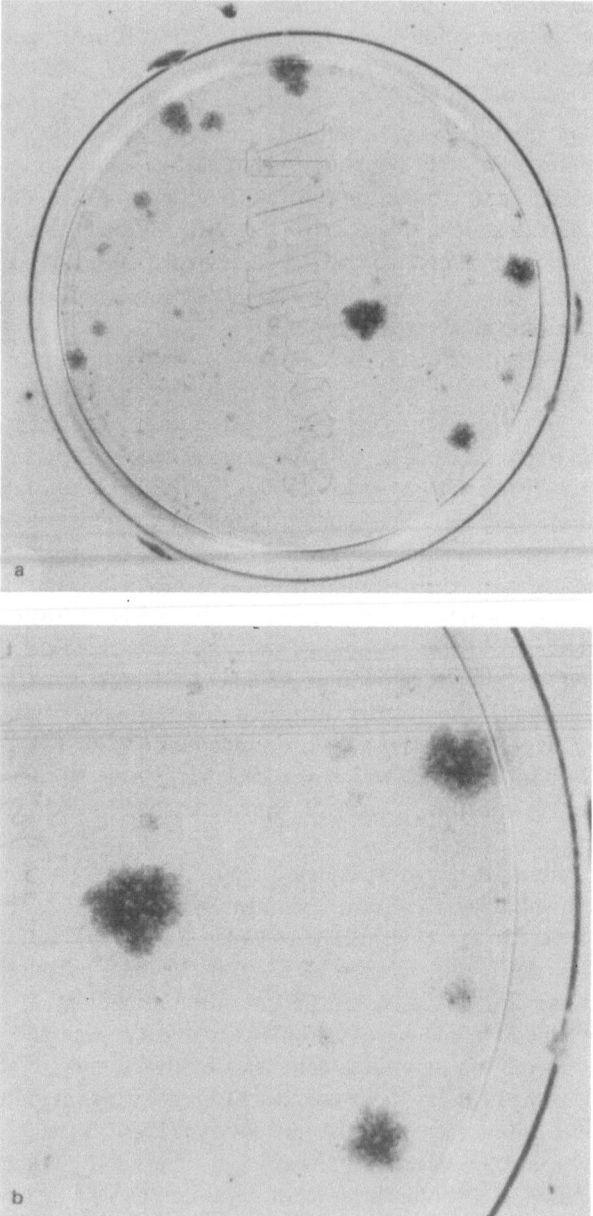

Fig. 30. Colonies of fibroblasts in monolayer cultures of peripheral blood leukocytes.

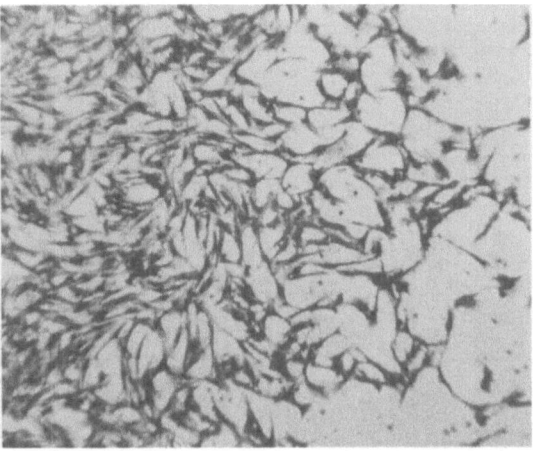

Fig. 31. Composition of a single colony in a 12-day culture of peripheral blood cells.

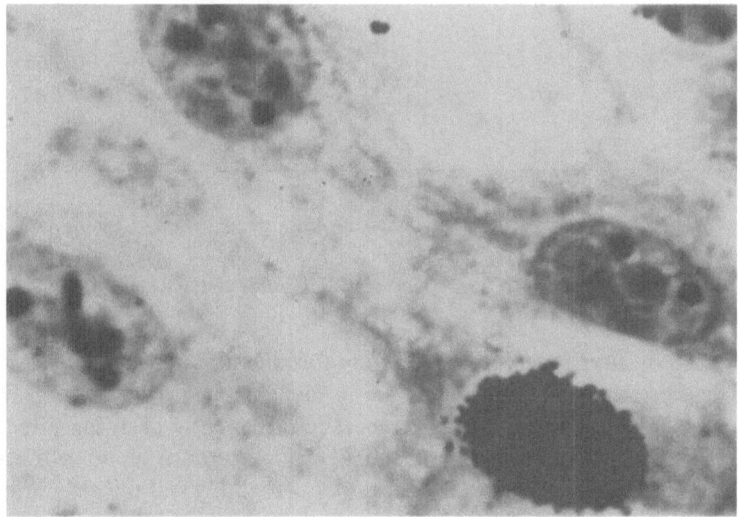

Fig. 32. Incorporation of thymidine-H³ into cells of colonies in peripheral blood cultures.

TABLE 10. Formation of Colonies of Fibroblasts in Monolayer Cultures of Blood Leukocytes[a]

Donor	Number of explanted cells	Number of colonies	ECF[b] per 10^5 cells
1	7.7	99	1.3
2	6	64	1.0
3	14.5	303	2.0
4	8.8	175	2.0
5	9.0	69	0.7
6	15.4	33	0.2
7	3.5^c	64	1.7
7	3.5^c	68	2.0
7	3.5^d	63	1.9
7	3.5^d	66	2.0
8	3.5^d	103	3.0
8	3.5^c	108	3.0
8	3.5^c	104	3.0
9	13.0^c	24	0.2
9	13.0^d	34	0.25
10	7.7^d	26	0.3
10	7.7^c	37	0.4
11	10.0^c	23	0.2
11	10.0^d	29	0.3
12	1	2	0.2
12	2	7	0.35
12	3	9	0.3
13	1	15	1.6
13	2	54	2.7
13	3	60	2.0
14	1	3	0.3
14	1	3	0.3
14	2	6	0.3
14	2	5	0.25

[a]After Luriya *et al.* (1971).
[b]Efficiency of colony formation, showing how many colony-forming cells are present in 10^5 explanted cells.
[c]Blood obtained by single puncture of the heart.
[d]Blood obtained by two or three successive punctures of the heart.

blood does not alter the size of the colony-forming unit (Table 10). This finding shows that precursors of fibroblasts are in fact present in the peripheral blood and are not transferred to it by puncture, as has frequently been suggested.

Addition of thymidine-H^3 to 7- and 11-day cultures of blood cells provided a means of determining the proliferating pool (Fig. 32) and the generation time of fibroblasts in colonies. The saturation curves were used for this purpose, with the assumption that growth of the cell population is logarithmic. The labeling index of the fibroblasts over a period of 15 min, which reflects the percentage of cells in the S-period, was 18%. The proliferating pool for 7- and 11-day cultures was not less than 84 and 79%, respectively. It follows from these results that the mean generation time (T) for fibroblasts in 7-day cultures is not less than 34.5 hr and not more than 39.3 hr, while in 11-day cultures, it is not less than 36 hr and not more than 33.4 hr (Luriya *et al.*, 1971*a*). It is interesting to note that similar parameters were obtained for fibroblasts in cultures of guinea pig bone marrow (Keilis-Borok *et al.*, 1971).

By means of monolayer cultures, the number of precursors of colonies of fibroblasts can thus be determined in various populations of lymphoid cells. Since in every case the concentration of precursor cells is extremely low, it can be extremely difficult to find them by morphological methods in the original material, and new experimental techniques will have to be developed.

It is likewise not clear whether colony-forming precursor cells pass through the morphological stage of the histiocyte–macrophage when explanted into culture. That fibroblast formation is observed virtually without the stage of growth of a histiocyte population in cultures of thymus and lymph node cells and of lymphocytes from the thoracic duct (W. Bloom, 1927, 1928) makes it even more unlikely that the histiocyte–macrophage is a stage in fibroblast formation.

3. Suspension Cultures of Lymphoid Tissue and Peripheral Blood Cells

During cultivation of peripheral blood leukocytes from which the red cells have been removed by precipitation with gelatin, rapid degeneration of the lymphoid cells takes place. Leukocytes treated with phytohemagglutinin (PHA) behave differently. Among lymphocytes in these cultures suspended in the medium and adherent to the slide, some are transformed into large basophilic cells of lymphoblast type, which can undergo two or three divisions. It has not yet proved possible to obtain long-term cultures, and their life is limited to 80 hr (Harris and Littleton, 1966). PHA induces blast-transformation of lymphocytes from human peripheral blood (Rabinowitz, 1963), lymph glands, and tonsils (Winkelstein and Craddock, 1965; Oettgen *et al.*, 1966). In cultures of thymocytes, only some cells, derived mainly from the medulla of the thymus, react to PHA. It seems probable that these cells are mature, immunologically competent cells capable of migrating from the thymus into other lymphoid organs. Small lymphocytes have been shown to undergo transformation, whereas their less mature precursors (large and medium lymphocytes) are insensitive to the action of PHA (Metcalf and Osmond, 1966; Rieke, 1966). The suggestion has been made that only mature immunocompetent lymphoid cells can react to PHA (Weber, 1966; Fridenshtein and Chertkov, 1969).

As was later shown, not only PHA (saline extract of kidney bean), but also several other agents (extracts of American pokeweed, fenugreek, forage broad beans, and lentils; staphylococcal endotoxin and streptolysin), act similarly on lymphocyte cultures. During blast-transformation for 30 hr, the cells gradually increase in size, their cytoplasm becomes basophilic, their nuclei enlarge, and nucleoli are clearly distinguishable. At the same time, mitoses can be found, and their number reaches a maximum at the peak of lymphoblast formation. All intermediate forms from lymphocyte to lymphoblast can be observed in the cultures. Electron-microscopic analysis of the transformed cells shows that the number of free ribosomes and mitochondria in them is increased, lysosome-like inclusions and vesicles of the smooth endoplasmic reticulum are formed, and the Golgi apparatus develops intensively.

Synthetic activity is increased in the transformed cells. An increase in RNA synthesis is apparent 15 min after the addition of PHA to the culture, and a stronger effect is observed after 2 hr. RNA synthesis in lymphoblasts is DNA-dependent, for it is blocked by actinomycin D and is evidently linked with the formation of new messenger RNA. Protein synthesis increases at the same time as RNA synthesis. Only after an increase in the synthesis of m-RNA and protein in the lymphocytes, as a result of which the cell can pass through the process of transformation into a lymphoblast, does DNA synthesis begin. This does not happen before 24–36 hr of cultivation. Usually, up to 70–95% of the small lymphocytes in the cultured population undergo blast-transformation. The mechanism of action of PHA is not yet known.

According to Rogers (1976), PHA-stimulated human peripheral blood lymphocytes *in vitro* synthesize DNA that is excreted into the culture medium. While resting lymphocyte DNA contains one or two copies of sequences similar to excreted DNA per haploid genome, stimulated lymphocytes on days 3 and 4 of culture contain 3- to 4-fold more copies. Thus, PHA induces lymphocytes to selectively replicate several copies of a limited portion of their genome, copies that are then excreted into the culture medium.

Important factors in determining the behavior of cells in suspension cultures are the depth of the layer of nutrient medium and the cell density. The role of these factors is easily verified. Various modifications of the suspension culture method enable the cells to be kept in constant motion, either by rotating the tubes, stirring or shaking the cell suspension, or passing a stream of gas through the medium (Osgood and Muscovitz, 1936; Clemmesen *et al.*, 1948; Ruddle *et al.*, 1958). The cells under these conditions are in a suspended state and do not adhere to the surface of the slide. In some cases, constant renewal of both liquid and gaseous phases is provided for.

In the course of the experiment, samples of medium with cells suspended in it can be taken and films prepared for cytological analysis.

In most cases, the suspension culture medium is used for quantitative study of cell metabolism and for biochemical investigations requiring large quantities of material; it is also used in oncology to obtain stable lines of leukemic blood and bone marrow. More recently, the use of this method has yielded important data on antibody formation *in vitro* (see p. 123).

According to Moore (1975), there is at least one lymphoid cell in every 25,000 peripheral blood leukocytes that has the capability of surviving, adapting itself to culture conditions, and multiplying at a rate that eventually produces a self-sustaining cell line. Fresh cultures of blood cells contain many metabolically active cells, and after 3 days in culture, there is usually an increase in the number of mononuclear cells. Yet by 7–10 days, there are only a few scattered macrophage-like cells attached to the glass, and the viability of suspended cells is usually less than 25%. But after an average lapse period of 40–60 days, there is a rapid and sustained growth rate of these residual cells. The dramatic onset of the establishment of most lymphoid cell lines is signaled by the appearance of large clumps of cells in the medium and signs of rapid increase in metabolic activity. The subsequent loss of such cultures through failure of sustained growth should be less than 5%.

The explanatory theories for the prolonged lapse period include: (1) a mutation of a surviving cell, followed by rapid growth; (2) a malignant transformation of a surviving cell; (3) viral transformation either to a dedifferentiated state or to a malignant form; (4) nongenetic stimulation by antigens, chemicals, or cell injury; and (5) slow clonal growth of a single cell or of several cells until they attain a population that will support rapid growth by the release of unidentified cell products.

In contrast to the hundreds of lymphoid cell lines with B-cell characteristics established from normal humans and from patients with various hematopoietic malignancies, only a few cell lines have I-cell characteristics.

The predominant morphological form in lymphoid cell lines is the immature lymphoblast. Most cell lines contain very few mature lymphocytes.

Other cell types include monocytoid cells, macrophage-like cells with phagocytized cells and cell fragments, plasmablast-like cells, and multinucleated cells. The peculiar absence of fibroblast-like cells from established lymphoid cell lines demonstrates once more that lymphoid cells do not transform themselves into mechanocytes.

4. Organ Cultures of Lymphoid Tissue

With the proper choice of feeder and nutrient medium, the organ culture technique enables lymphoid tissue to be grown for a long period *in vitro* while maintaining, or even regenerating, its structures and maintaining its normal histogenesis. These properties have been demonstrated (Luriya, 1966) using the multiple organ culture method.

Thymus

During culture of the neonatal mouse thymus in a Conway dish on HA millipore filters (pore size 0.45 µm) on nutrient medium consisting of 75% medium 199, 20% inactivated bovine serum, 5% chick embryo extract, glucose, vitamin C, and antibiotics, intensive migration of thymocytes takes place, so that after 48 hr of cultivation, they form around the explanted fragment a wide zone that is 3–4 times greater in diameter than the explant itself (Luriya, 1966). The migrating lymphocytes degenerate rapidly. Degeneration also spreads to the central zones of the explant. Only the periphery of the fragment remains viable, where intact lymphocytes are diffusely distributed among the large stromal cells, which form a wide-looped network. In the regions of preserved lymphocytes in the peripheral zone, appreciable numbers of mast cells are also present. The explant itself becomes epithelized: it is covered with a single or double layer of epithelial cells (Fig. 33) from which bands of cells and their branches, forming small cysts, penetrate into the interior of the fragment.

These changes are followed by active phagocytosis and digestion of the debris. In parts of the stroma freed from debris in the peripheral zones of the explant, small clusters of small and medium lymphocytes appear, as well as groups of large pyroninophilic cells of the lymphoblast type, many of which divide, persist, and gradually increase in size. In places, the small lymphocytes are densely packed in the lymphatics of the explants (Fig. 34). The connection between the mast cells and foci of lymphocyte proliferation is very clearly distinguishable at this stage.

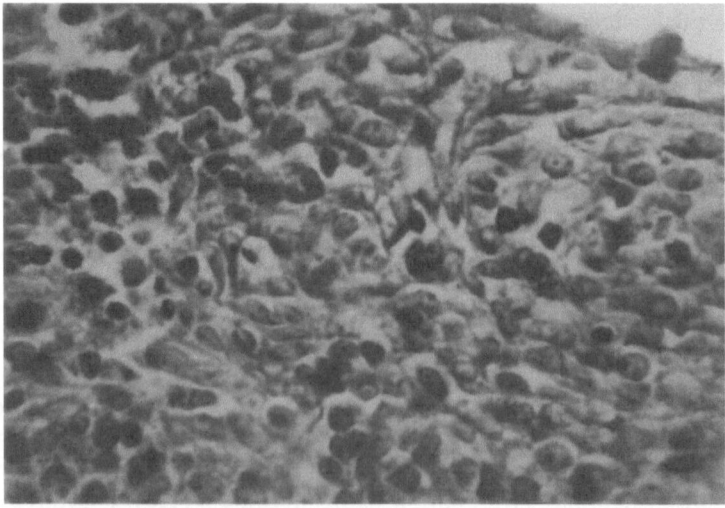

Fig. 33. Proliferation of the stroma in a 5-day organ culture of the thymus.

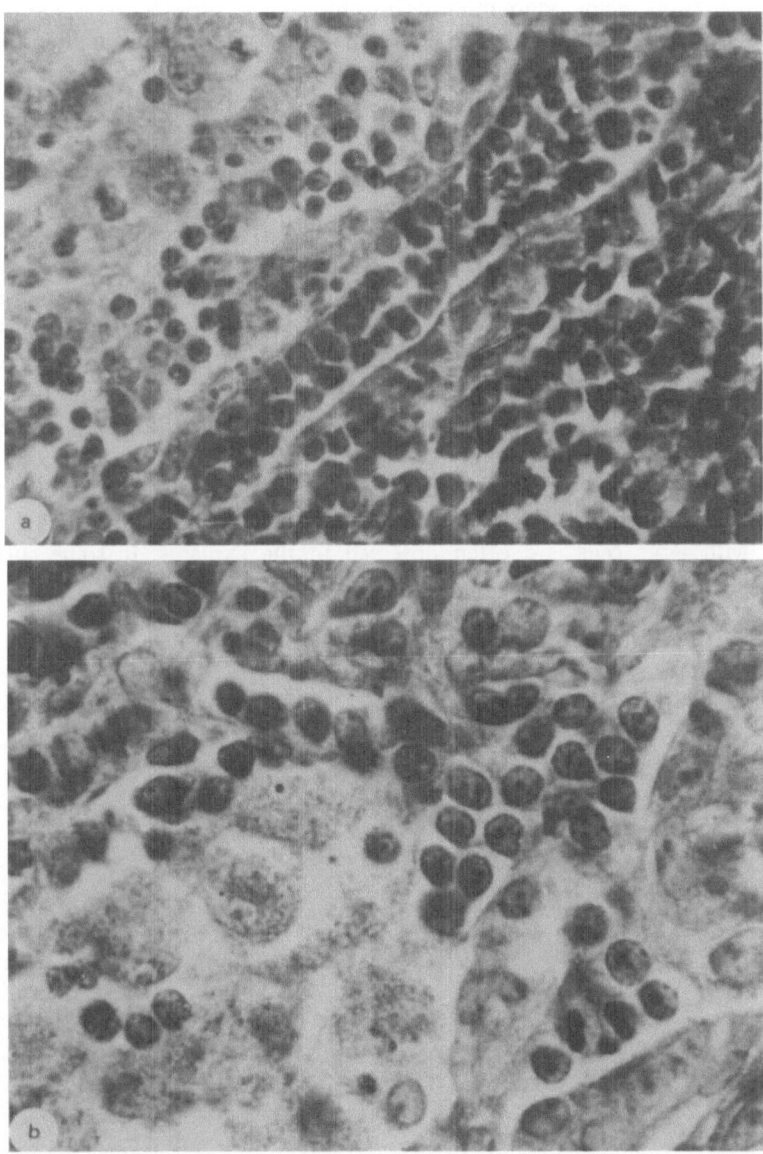

Fig. 34. Concentration of lymphocytes in lumen of lymphatics in 5-day organ cultures of the thymus.

The filter is also cleared of debris. On it, pictures indicating intensive proliferation of the stroma can be observed: proliferation of bands and sheets of cells. Migration of the lymphoid cells into the zone of growth ceases. On the 4th or 5th day, small foci of myeloid hematopoiesis are frequently found in thymus explants.

Regeneration of lymphocytes at the periphery of the explants leads to the formation of small follicles, clearly distinguishable against the background of the depopulated stroma. The follicle contains lymphoid cells at different stages of maturity and cells in a state of mitosis.

Later (10–12 days), the follicles increase in size, and lymphoid tissue fills the flattened explant. The thymus then seems to consist not of a cortex and medulla, but of separate round follicles (Fig. 35), frequently with translucent centers. Intensive proliferation of the stroma continues on the filter. The edges of the zone of growth consist of fibroblast-like cells forming grass-like out-growths. The lymphoid tissue, however, does not spread outside the fragment: only solitary lymphocytes are found in the zone of growth in the immediate vicinity of the explant.

In cultures aged 15–25 days, the fragment is filled, as before, with lympho-cytes that are distributed either as follicles or diffusely. Large epithelial cysts, lined with large cells, are often observed in the explants. In some cultures,

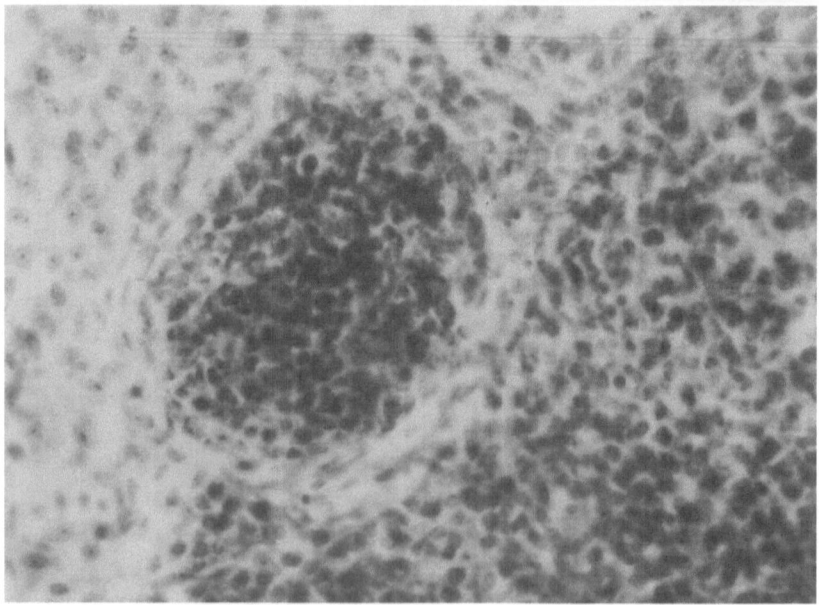

Fig. 35. Secondary follicle in 11-day organ culture of the neonatal mouse thymus.

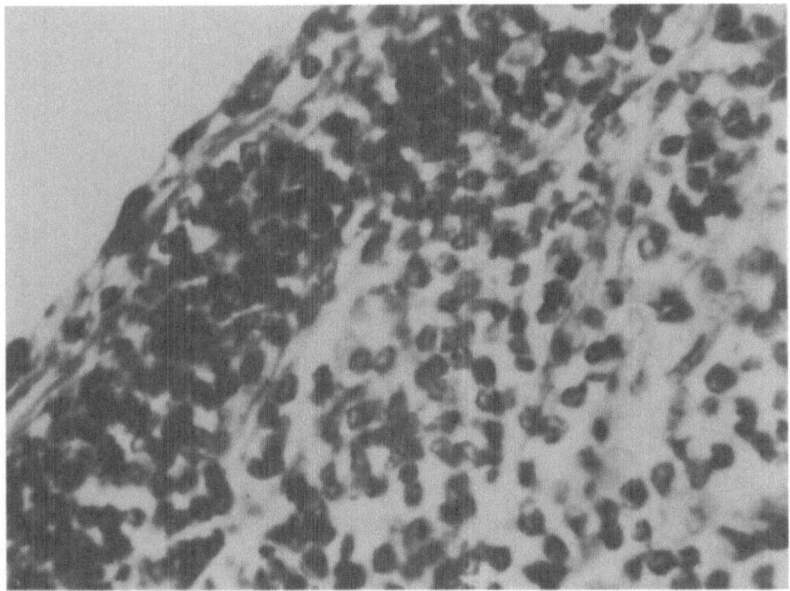

Fig. 42. Groups of lymphocytes under the capsule in a 5-day organ culture of a lymph node.

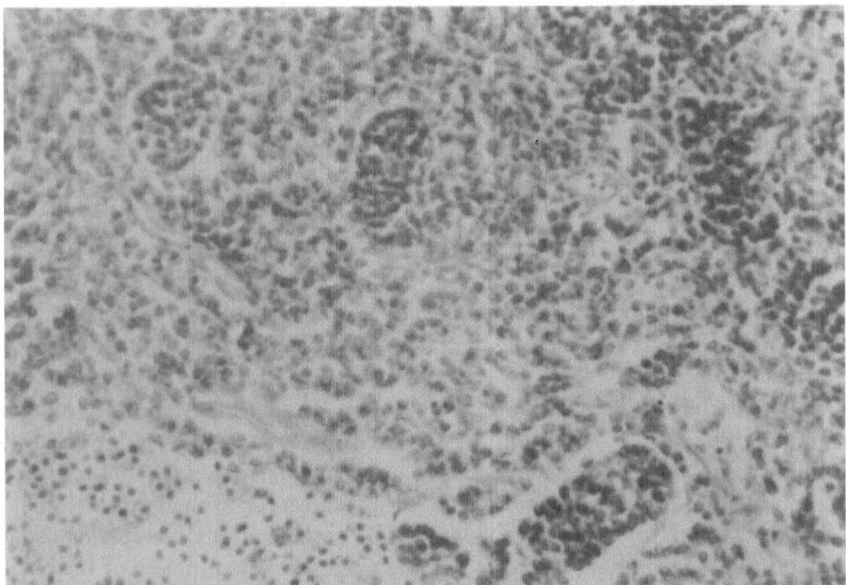

Fig. 43. Proliferation of lymphocytes in a 5-day organ culture of a lymph node.

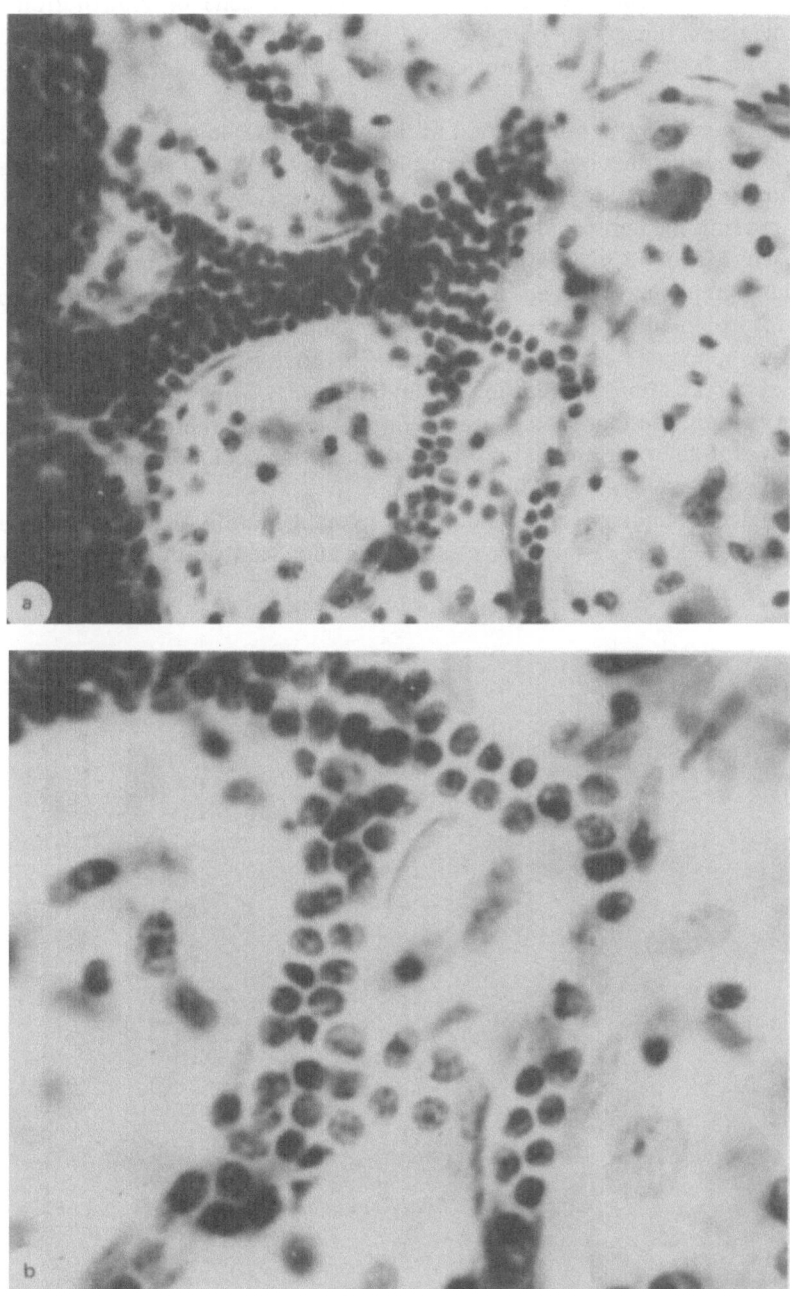

Fig. 36. Structures resembling blood vessels in a 15-day organ culture of embryonic thymus.

structures resembling blood vessels, entering the zone of growth from the fragment and filled with lymphocytes, are formed (Fig. 36). In cases in which these vascular structures do not appear, lymphocytes do not migrate from the explant.

Evidence of active proliferation of the lymphocytes in organ cultures is given not only by the presence of mitoses in the lymphoid cells, but also by data on the incorporation of thymidine-H^3 (Fig. 37). When thymidine-H^3 was added to the nutrient medium of 10-day cultures of the mouse neonatal thymus for between 1 and 72 hr, it was found that the proliferating pool for large and medium lymphocytes exceeds 50% (Table 11). After labeling for 1 hr, thymidine was incorporated into 3.5–5% of the large and medium lymphocytes; all the small lymphocytes remained unlabeled. The first labeled small lymphocytes appeared 7 hr after the addition of thymidine-H^3. These results show that in organ culture, the same categories of lymphocytes can proliferate as *in vivo*, and the time of their differentiation from the large or medium lymphocyte to the small lymphocyte is about 7 hr.

Generally similar results are obtained by explantation of thymus fragments of young guinea pigs. Initially, massive migration and death of the lymphoid

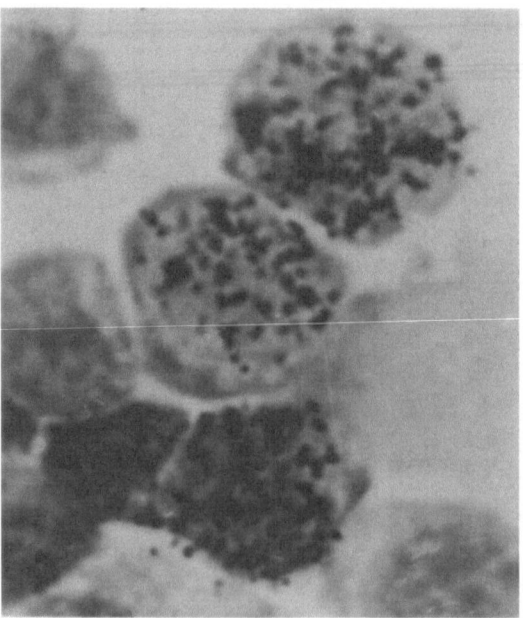

Fig. 37. Incorporation of thymidine-H^3 into large lymphocytes in a 10-day organ culture of the thymus.

TABLE 11. Changes in Labeling Index among Large–Medium and Small Lymphocytes on the Addition of Thymidine-H^3 to a 10-day Culture of Neonatal Mouse Thymus for 1, 3, 7, 24, 48, and 72 Hours[a]

Categories of lymphocytes	Formula (%)	Labeling index (%)					
		1 hr	3 hr	7 hr	24 hr	48 hr	72 hr
Large–medium lymphocytes	88	3.5– 5	6	7	16.5– 27	42	53
Small lymphocytes	12	0	0	0.4	7	11	18

[a]After Prusevich and Luriya (1972).

cells take place, leaving only a few of these cells in the peripheral areas of the explant. Later, the stroma is cleared of debris, it spreads over the filter, and lymphoid tissue in the explant itself regenerates. It is interesting to note that epithelial structures resembling Hassall's corpuscles are formed in cultures of guinea pig thymus. The cells forming them proliferate actively, as is shown by their intensive incorporation of thymidine-H^3. Compact groups of lymphocytes associated with cells of the stroma persist for a long time (30 days).

Fragments of guinea pig thymus were grown on the same nutrient medium as the mouse thymus, but the millipore filters differed in certain characteristics. In the first case, HA filters with a pore size of 0.45 μm were used; in the second case, RAWG and AUFS filters with pore sizes of 1.2 μm and 0.6–0.9 μm, respectively, were used. HA filters are impermeable to cells, while cells can pass freely through AUFS and RAWG filters. These types of filters probably also differ in their surface properties, a matter of great importance for the tissue (Fig. 38). As will be clear from the following description, millipore filters with a relatively large pore size are best used for long-term culture of lymphoid organs.

Lymph Nodes

If whole mouse and guinea pig lymph nodes are grown on HA and AUFS filters (Luriya and P'yanchenko, 1967), total destruction of the lymphoid cells accompanied by preservation of certain cells of the stroma takes place in the center of the explant in the early period (2–4 days). Small groups of living lymphocytes persist at the periphery of the explant. Sometimes collections of lymphocytes lie immediately beneath the capsule in dilated sinuses and in the lumen of lymphatics (Fig. 39). Large pyroninophilic cells are also found here.

In parts with a smaller cross-sectional area, the region of destruction of the lymphocytes is reduced, and the zone of their preservation and periphery is wider.

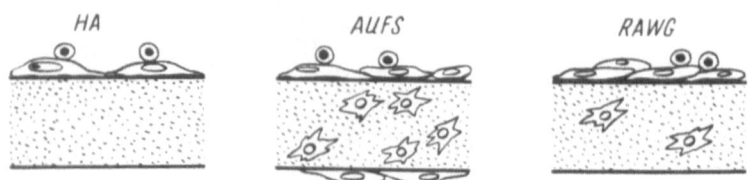

Fig. 38. Type of growth of cells in relation to structure and pore size of filters.

A circular zone of growth consisting of connective tissue cells surrounds the explants. When AUFS filters are used, branched connective tissue cells invade the pores of the filter and form a stratified zone of growth. Large fibroblasts on HA filters form a network in the loops of which lymphocytes can be seen singly and in groups (Fig. 40). Often the lymphocytes are in direct contact with the

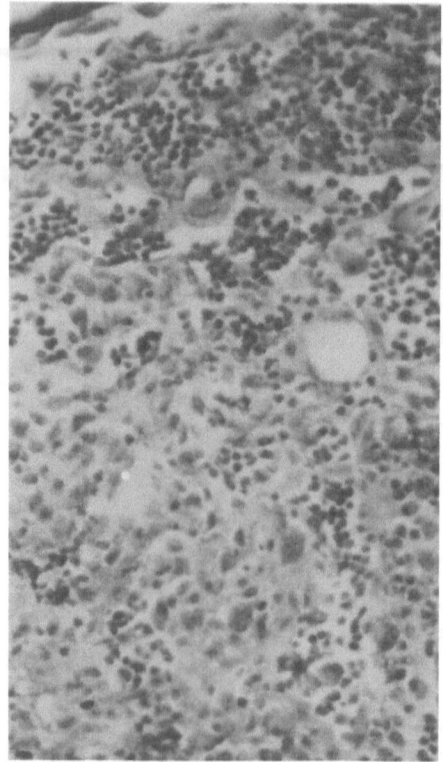

Fig. 39. Group of lymphocytes beneath the capsule and in the lumen of lymphatics in a 4-day lymph node culture.

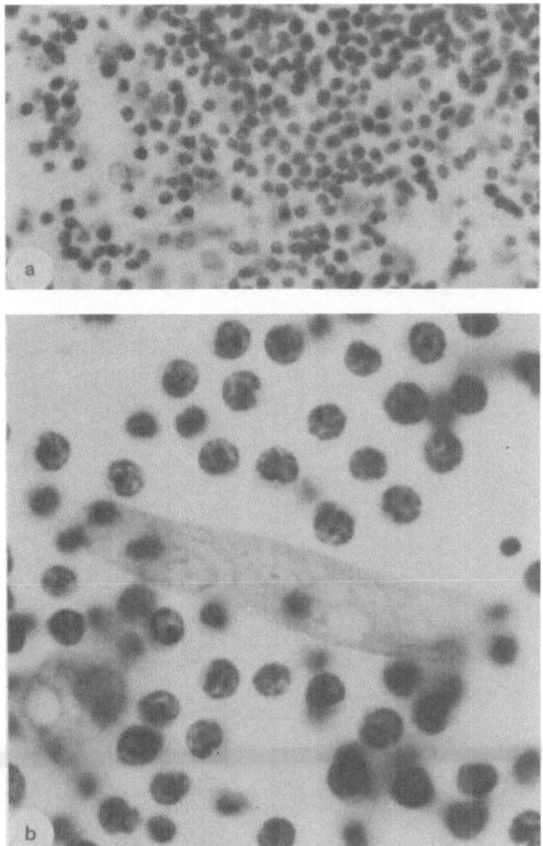

Fig. 40. Lymphocytes and reticulum cells on a filter in an 8-day organ culture of a lymph node.

cytoplasm of the stromal cells and with large cells resembling macrophages, phagocytosing degenerating lymphocytes.

As a rule, the line marking the front of emigrating lymphocytes follows that of the stromal cells. At the edge of the zone of growth, groups of cells consisting of lymphocytes closely surrounding a connective tissue cell (macrophage or fibroblast) can be seen on the filter (Fig. 41).

Later, the central parts of the explants are cleared of cell debris, and the stroma proliferates to fill the medulla of the lymph node. Lymphocytes are distributed as discrete foci among the macrophages of the stroma or in the vessels (Figs. 42 and 43) and are often associated with mast cells.

On the 11th or 12th day of growth, the explants consist of proliferating stroma occupying the central region, and of round groups of lymphocytes,

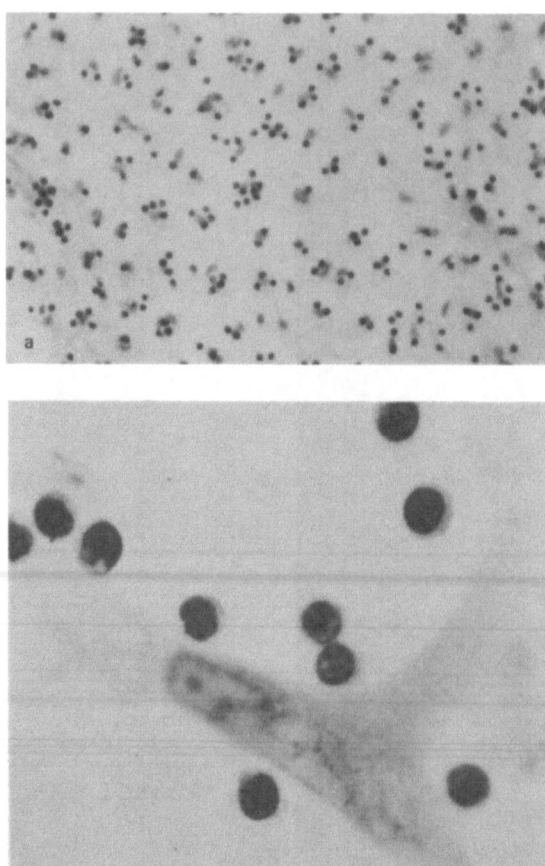

Fig. 41. "Feeder" reticulum cell surrounded by lymphocytes on filter in organ culture of a lymph node.

resembling follicles in shape, lying at the periphery of the lymph node. Each follicle contains several hundred lymphoid cells (Fig. 44). Solitary plasma cells are found among the lymphocytes.

During cultivation of lymph nodes on AUFS filters, the lymphoid tissue survives longer than on HA filters. By the end of the 3rd week of cultivation, aging of the stroma takes place in cultures grown on HA filters; this aging is expressed, in particular, as the accumulation of a brown pigment in the cells. The lymphoid tissue undergoes atrophy. The viable stroma and lymphoid tissue survive for more than a month in cultures of lymph nodes in AUFS filters. Even when these filters are used, however, no regeneration of the medulla takes place.

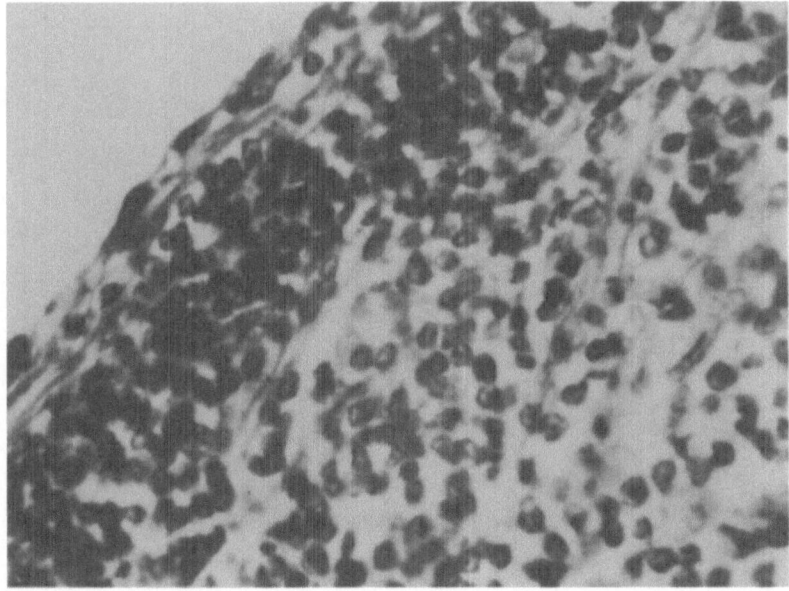

Fig. 42. Groups of lymphocytes under the capsule in a 5-day organ culture of a lymph node.

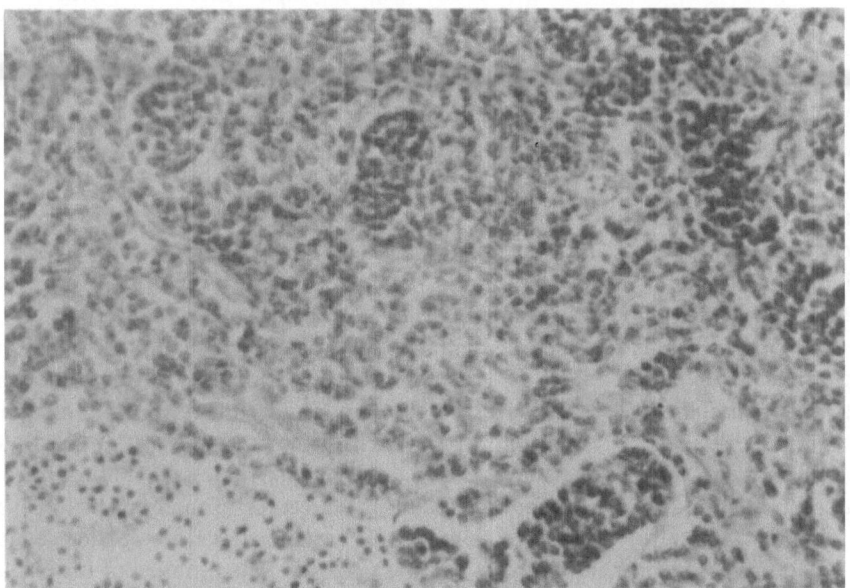

Fig. 43. Proliferation of lymphocytes in a 5-day organ culture of a lymph node.

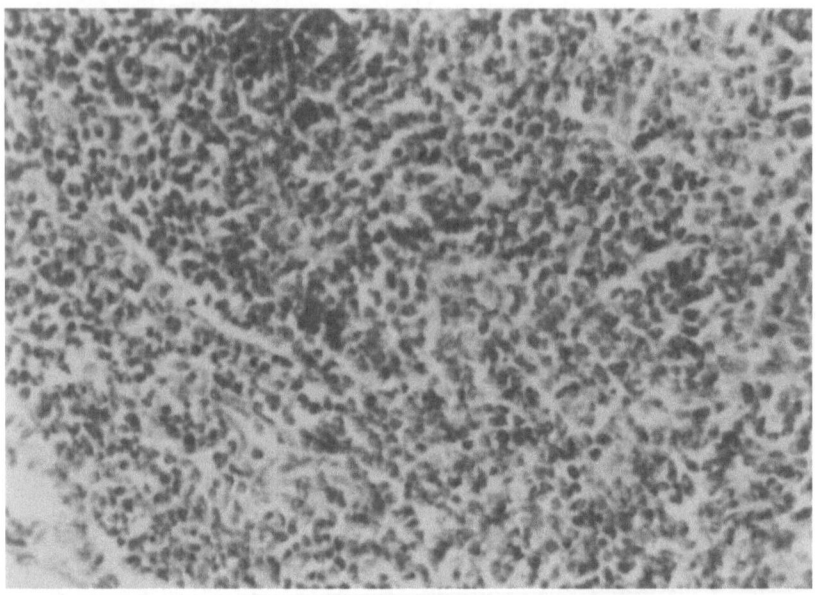

Fig. 44. Lymphoid tissue in a 15-day culture of a lymph node.

The ratio between the numbers of T and B lymphocytes was determined in 15-day cultures of lymph nodes from C3H mice. A cell suspension was prepared from the explants and used in the cytotoxic test with anti-theta C3H serum from immunized AKR mice. On incubation with normal mouse serum and after the subsequent addition of complement, the percentage of living cells was 86%, while if anti-theta serum and complement were used, the percentage of living cells was reduced to 47%. Since it is the T-lymphocytes that are destroyed by this reaction, the ratio between the T and B cells in the culture can be calculated. It was found that there were 55% B lymphocytes and 45% T lymphocytes; i.e., their distribution in organ cultures of lymph nodes is shifted toward the B lymphocytes by comparison with the normal lymph nodes.

In organ cultures of guinea pig lymph nodes grown on a medium with homologous serum, the same phases of destruction, regeneration, and restoration of the lymphoid structures of the cortex are found as in cultures on heterologous serum (Luriya and P'yanchenko, 1967). The distinguishing feature of these cultures is the structure of the regenerating secondary follicles. Spherical formations consisting of stromal cells are found in the center of many of them (Figs. 45 and 46). In culture, of course, migration of lymphocytes from outside cannot be the source of regeneration of the lymphoid tissue. The problem of the extent to which the reticulum cells of the stroma and lymphocytes surviving the period of

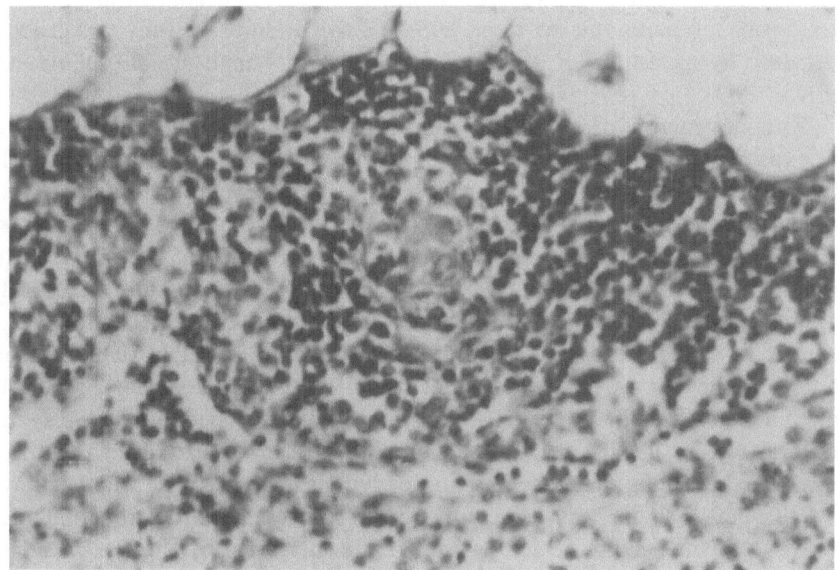

Fig. 45. Lymphoid follicle in the cortex of a lymph node with a spherical structure in the center. Time of cultivation, 8 days.

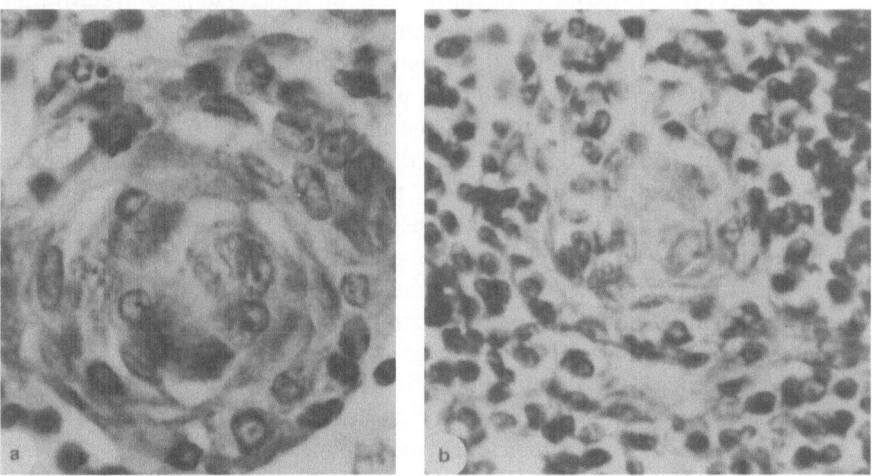

Fig. 46. Spherical structures in the center of lymphoid follicles.

destruction participate in this process can therefore be examined more clearly than during regeneration after transplantation. An attempt to study this problem was made by explantation of previously irradiated lymph nodes (Didukh and Luriya, 1967).

Total destruction of lymphocytes both in the center and at the periphery of the explant is observed in cultures of lymph nodes from irradiated (2000 R) mice. At the same time, living stromal cells were preserved in the explant; they were distributed separately among the degenerated lymphocytes. Phagocytosis of the cell debris by macrophages and gradual clearance of the stroma with encapsulation of the central region containing the debris took place. The lymphoid tissue, however, does not regenerate in these cultures. Although it is difficult to assess the degree to which the stromal cells remain viable in these cultures, if they are damaged, they evidently suffer to a far lesser degree than the lymphocytes. One therefore gains the impression that lymphocytes play the principal role in the regeneration of lymphoid tissue *in vitro*. That they do is shown by the results of transplantation of lymph nodes of mice irradiated at a dose of 2000 R. Lymphoid tissue in the grafts regenerated through repopulation of the recipient's lymphocytes on the irradiated stroma of the lymph nodes, which is consequently suitable for population by lymphocytes and for maintaining their differentiation.

The fate of the explant in organ cultures of lymphoid tissue is determined largely by the composition of the nutrient medium and the size of the explanted fragment. The smaller the fragment, the more limited the zone of necrosis that arises after explantation. If very small lymph nodes are explanted, the zone of necrosis may be virtually nonexistent.

The use of protein-free media has proved unpromising for long-term cultivation (Trowell, 1955, 1959). In these experiments, Trowell grew lymph nodes (whole or halved) on synthetic medium T8, T9 (without serum). A thin layer of 2% agar or a piece of lens paper was placed on a platform consisting of a metal grid in order to prevent contact between the tissue and the metal.

The histological study of sections of the cultures and cell suspensions from the explants showed that after 2 days, the proportion of pyknotic forms is 1–2%; after 4 days, 1–3%; and after 7 days, 20% of the cells. Not all cultures survive *in vitro* longer than 10 days, and many died sooner. The use of a synthetic medium is probably an unfavorable factor for growth of lymph nodes, and the deficient feeding leads to degeneration of the tissue.

Much better results are obtained when lymphoid organs are grown on medium containing serum and embryonic extract (Ball and Auerbach, 1960; Pinkel, 1964; Auerbach, 1965; Hannoun and Bussard, 1966; Luriya, 1966).

Conversion of the epitheliomesenchymal rudiment of the thymus of 12-day mouse embryos into a lymphoid organ containing large, medium, and small lymphocytes, as well as reticulum cells and epithelial stromal cells, has been demonstrated in organ culture on millipore filters (Auerbach, 1965). In a similar system, the spleens of mouse embryos undergo necrotic changes after cultivation for 1 week. In the case of combined explantation of spleen and thymus, proliferation and differentiation of the lymphocytes continue up to 5 weeks.

The behavior of lymphoid tissue in organ culture depends largely on the structure of the substrate supporting the explant. For instance, if platforms covered with lens paper are used, fragments of rabbit lymph nodes and spleen (Hannoun and Bussard, 1966) preserve their basic structures during the first days. Starting from the 4th day, the cell density in the fragment falls as a result of emigration of the cells. This process continues until the 30th–40th day of cultivation, after which only a depopulated stroma remains in the explant, sometimes with a zone of necrosis in the center. Lymphoid cells migrating from fragments fall down to the bottom of the culture vessel when the substrate for tissue growth is paper, which is a wide-looped network of interwoven fibers.

Fragments of human lymph nodes have also been successfully cultured on the surface of a gelatin sponge on medium F-10 with the addition of 10% neonatal calf serum (Kenneth, 1971). During the 1st week, migration of small lymphocytes and macrophages into the medium was observed. Unlike lymphocytes, the macrophages adhere to the bottom of the dish, and 3–4 weeks after explantation, fibroblast-like cells were found on the bottom. In the explant itself, moderate necrosis of the central regions was observed. At the periphery of the fragment, well-preserved lymphocyte and stromal cells remained. In the 4th week, lymphoid cells formed clusters at the periphery of the explant that gradually became larger, taking in more and more cells. The central part of the explant became fibrous in character. After the 2nd week of cultivation, the matrix of the sponge was invaded by fibroblast-like and lymphoid cells, and gradual development of the characteristic structural organization of the lymph node was observed. At the periphery of these structurally organized concentrations of stromal cells and lymphocytes, intensive proliferation of lymphoid cells, mainly lymphoblasts, took place during the 5th–9th weeks.

The most promising method of organ culture of lymphoid tissue at present seems to be culture on millipore filters. With this method, not only can proliferation and differentiation of the lymphoid cells be maintained for a long time, but also such complex processes as regeneration with the formation of the specific structures of lymphoid tissue can take place *in vitro*.

5. Comparison of Various Methods of Lymphoid Tissue Culture

The Behavior of Lymphocytes

It will be clear from what has been said that the behavior of lymphoid tissue *in vitro* is determined largely by the method of cultivation. The use of cultures in a plasma clot permits only temporary survival of lymphocytes. During the first days of cultivation, the explant loses the morphology of lymphoid tissue and comes to resemble connective tissue. The method of explantation in a plasma clot provides only limited opportunities for intravital observations of the cultures. For this purpose, it is more suitable to grow lymphocytes or fragments of lymphoid organs in special culture chambers that also permit microfilming (Gowans, 1957; Pulvertaft, 1959; Klein, 1958, 1959).

Lymphocytes remain intact for only a few days in chambers *in vitro*. Of lymphocytes from the rabbit thoracic duct, 80% survive for 24 hr, but only 20% for 48 hr (Gowans, 1957). With the use of a Jayne and Pulvertaft's chamber, in which the cells are placed in a drop of medium on nutrient agar and a slide placed on the top, lymphocytes survive up to 4 days (Pulvertaft, 1959); close contact between lymphocytes in the chamber and other cells results in their longer survival. The method of suspension culture of lymphoid cells likewise does not ensure their long survival. According to Trowell (1965), 15% of lymphocytes die during preparation of the suspension, 50–70% during cultivation for 24 hr, and up to 90% during the first 48 hr.

Substantial changes can be observed, however, in monolayer and suspension cultures in the presence of PHA or antigens (see below). Blast-transformation—i.e., the formation of large lymphocytes from small, and proliferation of lymphoblasts—takes place under the influence of PHA and other mitogens. Differentiation induced by PHA is of very short duration (on the order of a few days). It is associated with important and regular changes in the synthesis of nucleic acids and protein.

Prolonged histogenesis of lymphocytes is maintained in organ cultures, where it is associated with specialized structures of the lymphoid tissue, namely, secondary follicles, which persist and may even be formed *de novo* under these conditions. Histogenesis of plasma cells *in vitro* takes place with great difficulty even in organ cultures. The ratio between the number of plasma cells and lymphocytes in organ cultures of lymph nodes is much lower than normal. The reason may be that *in vitro* the structure of the medulla is preserved and regenerates to a much smaller degree than the structure of the lymph node cortex.

An interesting feature that distinguishes lymphoid tissue growing in organ culture is the presence of mast cells.

Proliferation of numerous macrophage—histiocytes and, in the late stages, of fibroblasts is invariably observed in cultures of lymphoid tissue. This applies equally to monolayer, plasma, and organ cultures. Admittedly, depending on the method of cultivation and the character of the explanted tissue, the number of histiocytes and fibroblasts that arise in the culture at the time of their appearance may vary.

Cell Transformations

The question of the origin of macrophages in cultures of lymphoid tissue is one that investigators have attempted to solve for more than 50 years. Even with respect to cultures of blood and lymph, which are cell populations that consist entirely of circulating cells, it is difficult to give a clear answer to the question of which cells (lymphocytes or monocytes) give rise to the macrophages. Monocytes account for only about 1% of the cells in lymph; for this reason, the dynamics of the appearance of macrophages *in vitro* and the small number of mitoses seemed to suggest that most macrophages arise from lymphocytes. It was concluded, however, that *in vivo*, macrophages arise from monocytes. The latter develop from bone marrow promonocytes (Van Furth and Cohn, 1968).

Accordingly, it has been shown that macrophages appearing in cultures of the spleen, thymus, and lymph nodes are the progeny of cells of bone marrow origin (Virolainen, 1968). With the use of tissues of radiation chimeras obtained by injection of syngenetic hematopoietic cells distinguished by a chromosomal marker $(T_6 T_6)$, the karyotype of macrophages grown in culture has been successfully analyzed. All macrophages obtained in cultures of the thymus, spleen, and lymph nodes had the chromosomal marker of the donor. The origin of the fibroblasts in lymphoid tissue cultures remains an unsolved problem.

As has already been stated, the formation of colonies of fibroblasts in monolayer cultures of blood and lymphoid organs does not agree with the assumption that the fibroblasts actually arise from histiocytes. The origin of fibroblasts in cultures of lymphoid tissue and blood is therefore a special problem that is closely connected with the origin of fibroblasts in bone marrow cultures (see p. 148).

The literature on cultivation of lymphoid tissue contains only a few references to the possibility of myeloid hematopoiesis in explants (Maksimov, 1917, 1923*b*; Latta and Johnson, 1934; Bloom, 1928).

For instance, in plasma cultures of lymph nodes containing bone marrow extract, Maksimov (1917, 1923*b*) observed cases in which myeloid cells and megakaryocytes appeared. Myeloid cells were observed on the 4th day of cultivation. Eosinophilic myelocytes were found to appear in cultures of mesenteric lymph nodes of rats (Latta and Johnson, 1934). Although many investigators were unable to reproduce myeloid hematopoiesis later in lymphoid tissue cultures, this problem merits the closest attention.

Small foci of myeloid cells, among which are some in a state of mitosis, also appear frequently in organ cultures of neonatal mouse thymus on the 4th or 5th day after explantation (Luriya, 1966). The myeloid cells lie beneath the explant in close contact with the stroma. As a rule, they lie separately from the groups of lymphocytes.

It is difficult to judge which morphological forms these hematopoietic cells arise from. The cases of hematopoiesis described above are evidence, however, that hematopoietic precursor cells are present in lymphoid organs, and that after explantation, they are able to realize their potential for differentiation. The factors responsible for this differentiation after explantation *in vitro* are unknown. It can only be postulated that when the normal packing of the cells is disturbed, the probability of deviation from the ordinary (lymphoid) pathway of histogenesis is increased.

It is useful to recall that under normal conditions, myeloid metaplasia of lymphoid organs is rarely observed *in vivo*. To judge from results obtained with irradiated mice, hematopoietic stem cells are found in negligible numbers in suspensions of lymph node and thymus cells. The suggestion has been made that they were found among the blood leukocytes contained in blood vessels within the lymph nodes and thymus. All these considerations make the discovery of foci of hematopoiesis in cultures of lymphoid organs particularly interesting.

The Process of Regeneration

Unlike other methods of explantation, the organ culture method provides favorable conditions for the preservation and proliferation of lymphocytes for a long time. Viable lymphocytes can be observed in organ cultures in the very late stages of cultivation (after 290 days) (Pinkel, 1964).

Explants of lymphoid tissue in organ cultures pass through a series of successive stages characterized by sharp differences in cell composition. During the first days *in vitro*, degeneration of most of the lymphocytes takes place, with only a few remaining at the periphery of the explants. Meanwhile, migration of the lymphocytes and their death en masse are observed on the filters. This process is particularly characteristic of cultures of the thymus. Viable lymphocytes persist in the zone of growth only where they are in contact with stromal cells. That the epithelial stroma of the thymus supports differentiation of lymphocytes has been demonstrated in experiments carried out by the diffusion chamber method *in vivo* (Luriya and Snegireva, 1966).

Starting from the 5th or 6th day of cultivation, regeneration of lymphoid tissue takes place in the cortical region of the explants. During both explantation of lymph nodes and cultivation of the thymus, regeneration takes place by the formation of lymphoid follicles, often with characteristic translucent centers. These structures are typical of lymph nodes, but are not found in the intact thymus *in vivo*, reflecting the different immunological rules of these organs. Antigens are known not to penetrate into the thymus, and antibody-forming cells do not differentiate in it. At the same time, after direct injection of an antigen into the thymus, secondary follicles are formed in it, and plasma cells appear.

A group of foreign antigens, contained in heterogenetic serum and chick embryo extract, is present in the culture medium used for explantation of the thymus. Presumably, it is through antigenic stimulation that regeneration of the lymphoid tissue proceeds in cultures via the formation of lymphoid follicles. In organ cultures of lymph nodes, regeneration of the cortical layer only is observed; the medulla does not regenerate.

Prolonged proliferation of lymphoid tissue and its regeneration *in vitro* can thus take place in organ cultures. What are the sources of this regeneration; i.e., through which cells (reticular cells or lymphocytes) does it take place? As pointed out already, explantation of lymph nodes from donors irradiated at a dose of 2000 R showed that lymphocytes are not formed from this irradiated stroma. One cannot rule out, however, the possibility that with such a high dose of irradiation, the properties of the reticular cells and their capacity for differentiation are altered.

Maksimov described the transformation of reticular cells into lymphocytes in plasma cultures of lymph nodes. He judged that transformation had taken place from the presence of the pigment characteristic of reticular cells in the lymphocytes. However, such a marker cannot nowadays be regarded as reliable. It can be concluded from morphological observations on organ cultures in successive periods that the chief (and probably the only) source of regeneration of lymphoid tissue is the viable lymphocytes remaining at the periphery of the explant. In groups of these cells on the 4th or 5th day of cultivation, mitoses are found, and later, secondary lymphoid follicles are formed in their place.

These observations agree well with data on the sources of regeneration of the lymphoid and stromal cells after transplantation *in vivo*.

Formation of Structures Characteristic of Interaction Between Lymphocytes and Stromal Cells

Although all methods of explantation other than organ cultures cannot maintain differentiation of lymphocytes for a long time, cases in which a very few lymphocytes in close contact with the stromal cells have survived have been described in several systems (Popov, 1927; Klein, 1958, 1959).

If lymphocytes were cultured together with embryonic heart muscle cells, with chick fibroblasts, or with human thyroid epithelium, a few lymphocytes survived for several weeks (Bloom, 1927, 1928; Pulvertaft, 1959).

One gains the impression that close contact between lymphocytes and other cells enables them to survive for long periods *in vitro*. Lymphocytes have a small amount of cytoplasm and very few mitochondria. The total volume of the mitochondria in the cytoplasm of the small lymphocyte is $0.5-1.3 \ \mu m^2$, and in the liver cell, $800 \ \mu m^2$ (Trowell, 1965). It can therefore be postulated that lymphocytes, when placed in contact with other "feeder" cells, receive from them essential metabolic products and energy in the form of ATP.

Morphological observations show that this type of relationship does in fact exist. In Trowell's words, the lymphocyte behaves more like an "active parasite" for which the reticular and epithelial cells are "active feeders." Lymphocytes are described as giving off thin, needle-shaped threads or pseudopodia sometimes reaching 100 μm in length (Pappenheimer, 1913; De Bruyn, 1945; Pulvertaft and Jayne, 1953; Shelton and Rice, 1959). In cultures of the thymus and spleen of the human embryo, lymphocytes can form long processes and form connections with reticular cells (Klein, 1958).

In organ cultures of lymph glands on millipore filters, a frequent discovery in the zone of growth is characteristic groups consisting of a reticular cell or "feeder," surrounded by lymphocytes. Lymphocytes located on the filter but not in contact with reticulum cells quickly undergo degenerative changes.

A special type of connection between reticular cells and lymphocytes can be seen in the formation of secondary lymphoid follicles in organ cultures. In many cases, the lymphoid follicle is formed around characteristic spherical structures formed by reticular cells.

The behavior of lymphoid tissue in organ cultures shows how important its interaction with the stromal cells is for the maintenance of differentiation of lymphocytes *in vitro*. This problem deserves careful study. One approach to its analysis may be by combining in one organ culture lymphocytes with pure lines of stromal cells from monolayer cultures of the thymus, spleen, lymph nodes, peripheral blood, and peritoneal exudate. From such experiments, information on the specific features of various types of stroma of lymphoid tissue and on their effect on differentiation of lymphocytes can be obtained.

Immunological Functions of Lymphoid Tissue *In Vitro*

1. The Process of Antibody Formation

Basic Mechanisms of Antibody Formation In Vivo and Aims of Research into This Process In Vitro

One of the chief functions of lymphoid tissue in the body is to carry out immunological reactions. This function is manifested primarily as the production of antibodies (immunoglobulins complementary to antigens) and as delayed hypersensitivity reactions.

General changes in lymphoid tissue, the participation of individual cells in antibody formation, and the role of cell structures in this process have been analyzed in detail (Nossal, 1966; Feldman and Bleiberg, 1967; Katz and Benacerraf, 1972; Unanue, 1972).

On the basis of recent experimental research, the following scheme of the successive events taking place in lymphoid tissue during antibody formation has been formulated. The injected antigen is phagocytosed by macrophages and reticular cells of the stroma of the lymphoid organs. In particular, the reticular tissue of the secondary follicles of the lymph nodes possesses increased phagocytic activity and increased ability to accumulate foreign antigens. The phagocytosed antigen undergoes further treatment in the reticular cells and macrophages. The next step is the transmission of the treated antigen (possibly in the form of a complex with RNA) to cells that act as original elements for the formation of clones of immune plasma cells.

The development of a clone of antibody-producing cells is preceded by mutual interaction among cells of the lymphocyte population itself. Lymphocytes of bone marrow origin (B cells), which are located mainly on the territory of the secondary follicles of the lymph nodes and spleen, have been shown to be precursors for the formation of a plasma cell clone. Activation of lymphocytes of bone marrow origin requires their interaction with lymphocytes derived from the thymus (T cells), which are found in the paracortical (thymus-dependent) areas of spleen and lymph nodes. The B lymphocytes enter the stroma of lymphoid organs (lymph nodes and spleen) directly from the bone marrow, while the T lymphocytes are formed in the thymus by differentiation of a polypotent stem cell under the influence of the microenvironment created by the stroma of the thymus. In its conversion in the thymocyte, the stem cell passes through a series of divisions and acquires new properties: the amount of H-2 antigens on its surface is reduced, and several specific thymic antigens appear on it (Schlesinger, 1972). The conditions of the microenvironment of the thymus, it must be emphasized, are such that types of genetically determined differentiation that cannot be followed by the stem cell outside the thymus can be exhibited within it. The maturing thymocyte (T lymphocyte) entering the circulation and repopulating the lymph nodes and spleen also has several antigenic and functional differences from thymocytes inside the thymus.

The formation of the clone of antibody-producing cells is preceded by a phase of activation by antigen of antigen-sensitive T cells that are not themselves precursors of plasma cells, but that stimulate or suppress the differentiation of B cells to actively secreting antibody-forming cells (J. F. A. Miller *et al.*, 1971).

For the immune response to take place, it is usually necessary to have not only the participation of macrophages and lymphocytes, but also cooperative interaction among the lymphocytes themselves.

Having once set out along the path of histogenesis of immune plasma cells, the precursor cell forms a clone of plasma cells synthesizing antibodies complementary to the inducer antigen. In the course of development of the clone, eight or nine successive mitoses take place. This repeated mitosis provides the conditions for the formation of a large population of antibody-synthesizing cells from one precursor cell. To begin with, cells of the large lymphocyte type (hemocytoblasts) appear, followed by proliferating plasmablasts and immature plasma cells, already capable of synthesizing antibodies, and, finally, mature plasma cells. These plasma cells are unicellular protein glands that produce immunoglobulins and are incapable of further multiplication; i.e., they are the final stage of differentiation. Different stages of histogenesis of plasma cells in antibody response take place in various parts of the lymphoid organs. Processing of the antigen takes place chiefly in reticular cells and macrophages of secondary follicles of the lymph nodes or stroma of the Malpighian bodies of the spleen, which are homologous structures. The degree to which these structures of the lymphoid organs participate in the immunological reaction depends on the method of immunization, i.e., the method of injection of the antigen. It is here that differentiation of the precursor cells, under the action of antigen and of T-helper cells, as well as the first stages of development of the clones of immune cells, take place. This is manifested morphologically as proliferation of large lymphocytes and an increase in the mitotic index. The more highly differentiated forms (plasmablasts and immature plasma cells), however, are not yet associated with the secondary follicles; they occur within them only rarely. These cells, as well as mature plasma cells, are found in the medullary cords of the lymph nodes or in the red pulp of the spleen.

Maturation of the clone takes about 6 days, while the life span of the mature plasma cells that synthesize antibodies is 8–48 hr. The scheme of the change in composition of the cell population during antibody synthesis evidently reflects only the most general rule. Although this scheme is in good agreement with curves of serum antibody titers, it evidently fails to take into account many important details of the differentiation of the antibody-synthesizing cells. In the course of histogenesis, not all cells of the clone in fact mature synchronously. In particular, after immunization, some mature plasma cells that synthesize antibodies appear much earlier than the time required for 8–9 cell divisions (Sainte-Marie, 1966). In addition, among the mature plasma cells, there are a few with a life cycle much longer than 48 hr.

All these data reflect our lack of knowledge of the cellular basis even of antibody synthesis, the most completely studied immunological phenomenon, which is based on development of clones of plasma cells. This lack applies even more to the formation of the original cells for these clones, i.e., to the interaction among macrophages, reticular cells, T cells, and B cells that forms the basis of this process.

These remarks also apply to the mechanisms of formation of the immulological memory, i.e., the appearance in the lymphoid tissue of a special category of small lymphocytes that ensure its preparedness for the secondary type of response to repeated injection of the antigen.

The process of antibody formation is associated with proliferation and differentiation of cells and with interaction among cells in specific structures of the lymphoid tissue. The role of the structures in this process is particularly in doubt. It is not known whether antibody formation can develop from the beginning in cells that are not organized into structures, i.e., whether the structures are an essential component in the formation of the immune response or whether they have the role of regulating factors that determine the specific dynamics of the process.

It is often difficult to study this problem and other problems in the intact organism. Lymphoid tissue is formed on the repopulation principle, as is reflected, for example, in the constant recirculation of the lymphocytes, the dissemination of cells of the immunological memory from one lymphoid organ into other lymphoid organs, and so on.

It must also not be forgotten that important factors regulating the course of immunological reactions exert their action on lymphoid tissue. These effects of the corticosteroids, in particular, are well known. These two circumstances—the high level of repopulation of cells within the lymphoid system and its great sensitivity to the action of certain regulatory factors arising from other tissues of the body—constitute the chief obstacles to the study of many problems in immunity. For this reason, cultures of lymphoid tissue may provide a valuable experimental model in some cases. That they may is evident even in that the method of transplantation of dissociated lymphoid cells (whether free or in diffusion chambers) has provided a basis for obtaining much important information on the cellular basis of immunity. Nevertheless, this method is merely a stepping-stone to explantation.

Cell populations responsible for immunological reactions are highly hetero-geneous. They include cells with various immunological functions: those that identify an antigen, those that present it, helpers, and supressors. It is therefore necessary to be able to test each of these categories separately as well as study them as a unity in analyses of the cellular bases of immunity. This possibility is provided by *in vitro* cultures. Important results have already been yielded, one being the establishment of the role of A cells in immunological response.

Before further progress can be made in immunological research, it is extremely important to have isolated systems in which lymphoid tissue can survive and remain capable of maintaining its state of equilibrium. Systems in which the structural organization of the lymphoid tissue is preserved, and in which specific stages of immunological reactions can be reproduced in cell populations as small and homogeneous as possible, are needed.

The composition of the medium for tissue cultures is chosen arbitrarily, and it can contain high concentrations of substances (including antimetabolites, antigens, etc.) that cannot be produced *in vivo,* and the duration of their action on the lymphoid tissue can be regulated at will. In addition, lymphoid tissue in culture is isolated from the hormonal background of the organism and the influence of the nervous system and of other factors, so that the results of the *in vitro* investigation reveal the direct response of lymphoid tissue to particular factors. Despite the obvious advantages of the isolated system *in vitro,* however, no perfect model for the study of immunity outside the living organism has yet been developed. This is particularly true of the induction of the primary immune response *in vitro.* Stimulation of secondary antibody synthesis has proved to be an easier task.

Secondary Antibody Synthesis In Vitro

A secondary immune response has been obtained *in vitro* in many experi-ments using different explantation methods: cultures of lymphoid cell suspen-sions, cultivation of lymphoid tissue fragments in a plasma clot in roller tubes, and explantation into organ cultures. Bacterial and other protein antigens have been tested as inducers of antibody formation. Immunological and chemical methods of differing sensitivity have been used to detect antibodies synthesized *in vitro*. In some of these investigations, the cultures have also been subjected to morphological analysis.

A comparatively simple method that allows antibody production by individual lymphoid cells to be monitored during their short period of survival *in vitro* has been developed. The cells are taken from an animal in a state of antibody synthesis (Nossal and Lederberg, 1958; Nossal, 1958, 1959, 1960, 1966) and placed one at a time in microdroplets consisting of synthetic medium with serum; the drops are covered with mineral oil. Various *Salmonella* antigens can be used to immunize the donors. After incubation of the cells in the microdroplets for 3–4 hr, together with a few of the corresponding bacteria, antibody production can be estimated by observing the fate of the bacteria, e.g., their motility.

With this model, Nossal showed that cells synthesizing antibodies belong to a series of plasma cells characterized by an eccentric position of the nucleus and a high RNA concentration in the cytoplasm. Each plasma cell obtained from an animal immunized with two or more antigens at the same time is usually capable of synthesizing antibodies of only one type. Only some cells (less than 2%) synthesize antibodies against two antigens simultaneously. In the microdroplet, the plasma cell exhibits synthetic activity and maintains its viability for only a few hours.

Suspension of lymphoid cells from immune donors constitute a surviving cell population that continues to synthesize antibodies *in vitro* for some time. The life span of the lymphoid cells under these conditions is limited to a few days. Cell suspension models are used chiefly for the biochemical analysis of antibody formation and for studying the chemistry of antibodies (Gurvich *et al.*, 1964; Skvortsov and Gurvich, 1968).

Special chambers suggested for studying immunity *in vitro* have made it possible to obtain suspensions of lymphoid cells that survived longer and synthesized antibodies intensively (Steiner and Anker, 1956). The suspension is poured onto a cellophane membrane placed above the reservoir containing nutrient medium. With constant shaking, the metabolic products are removed and the nutrient medium renewed.

Cultivation of lymphoid tissue fragments from immune donors leads to the survival of certain lymphoid structures and enables a prolonged secondary response to be obtained *in vitro*. Growing of pieces of lymph nodes in a plasma clot in roller tubes provides the best conditions for antibody synthesis. Under optimal conditions, antibody synthesis has been observed for 4 weeks (Michaelides, 1957; Stravitsky, 1961; Michaelides and Coons, 1963; *et al.*). It is interesting to note that the phase of inhibition of antibody synthesis in cultures was less marked than *in vivo*.

The role of several factors in the development of the secondary immune response in cultures has been studied. The composition of the nutrient medium was shown to play the decisive role in antibody synthesis *in vitro*. The presence of optimal concentrations of certain amino acids in the medium is an essential condition for antibody formation in culture (Mountain, 1955; B. Wolf and Stavitsky, 1958). The addition of vitamins A, K, and B_{12} and tocopherol stimulates antibody synthesis, while the addition of purines and pyrimidines, lipids, and carbohydrates has no such stimulating effect (Wolf and Stravitsky, 1958; Alfred *et al.*, 1963). In addition, the culture medium must contain serum. According to some observations, cortisone and some other corticosteroids, as well as insulin, can replace serum (Ambrose and Coons, 1963; Ambrose, 1964).

The secondary immune response has been investigated most thoroughly by Coons and co-workers. They studied the development of the secondary response in fragments of rabbit lymph nodes embedded in a plasma clot and placed in roller tubes. Bovine serum albumin (BSA) and diphtheria toxoid were used as the antigens. Lymph node fragments from rabbits in which an immunological memory had been formed were explanted into cultures. Reimmunization was carried out at the moment of transplantation or during the subsequent days of cultivation. Antibodies in the medium were determined by the passive hemagglutination test. The experiment showed that the lowest dose of BSA capable of inducing a secondary response *in vitro* is 1×10^{-9} g/ml (O'Brien *et al.*, 1963).

These workers showed that ability to develop a secondary response to diphtheria toxoid persists in the cultures for 4 days, and to BSA, for about 8 days. When both antigens are used for reimmunization *in vitro*, the responses proceed independently. The antibody titers reach a peak on the 5th–8th day after immunization. Antibody synthesis lasts about 4 weeks. The phase of inhibition of antibody formation in the cultures is less marked than *in vivo*. Most of the cells forming antibodies and detectable by Coons's immunoserological method remain in the fragment, although some are also found in the zone of growth (O'Brien *et al.*, 1963). Fluorescent cells begin to appear on the 3rd day. On the 4th day, these cells are widely distributed; most have the appearance of immature plasma cells, but on the 8th day, they are mainly mature plasma cells.

The action of 5-bromodeoxyuridine and chloramphenicol on the anamnestic response was studied *in vitro*. Addition of 5-bromodeoxyuridine (an inhibitor of DNA synthesis) during the first 3 or 4 days after immunization suppresses the secondary immune response *in vitro*, whereas addition of this agent to the medium after this period has no significant effect on antibody synthesis. The results of this investigation show that the development of the secondary response in cultures depends on cell division during the first 3 or 4 days after immunization (O'Brien and Coons, 1963).

Chloramphenicol (an inhibitor of RNA synthesis) suppressed antibody formation *in vitro*. Comparing the effects of treatment with chloramphenicol on the various stages of the immune response *in vitro*, it can be concluded that it inhibits antibody production by acting on one of the early phases of the immune response; i.e., one cannot rule out the possibility that inhibition of messenger RNA synthesis is responsible for this suppression (Ambrose and Coons, 1963).

Thus, some very significant results were obtained even by methods that resulted in the survival of lymphoid tissue *in vitro* for only a short time or in an obviously imperfect state.

With the use of organ cultures, lymphocytes can be made to survive and proliferate for a long time. In such cultures, it is evidently possible to expect a full secondary immune response.

In fact, fragments of lymph nodes from rabbits in which an immunological memory had been formed, when grown in organ cultures on paper rafts, showed considerable antibody synthesis on the 9th day, if the antigen was added to the medium at the moment of explantation. Antibody formation continued for about a month (Tao, 1964). It is interesting to note that the anamnestic response *in vitro* was induced in this system not only by the addition of antigen, but also by PHA. The proliferative activity of cultures receiving antigen or PHA (determined by incorporation of thymidine-H^3) was very high. Cells of hemocytoblast type, as well as mature and immature plasma cells, were found in the lymphoid tissue fragments.

Some interesting results were obtained by cultivation of lymph node and spleen fragments from rabbits immunized with sheep red blood cells (SRBC) in organ cultures (Hannoun and Bussard, 1966; Bussard and Hannoun, 1966). Two types of cells are liberated from the explants in the course of cultivation: free cells, which are in suspension and degenerate over a period of several days, and cells adherent to the glass, which retain their immunological activity for 3 weeks. Antibody production by individual cells was determined by the method of local hemolysis in gel. Several different cell types were found on the slide: large epithelioid cells, macrophages, fibroblasts, and blastoid cells. Cells adherent to the surface of the slide and giving a plaque of hemolysis were found to differ in their morphology; there were plasma cells that were flatter than normal and had probably changed their shape as a result of contact with the glass, together with reticulum cells with processes, the nuclei of which contained one or two nucleoli, and binuclear (or closely associated) cells.

A prolonged secondary response was obtained in organ cultures of fragments of lymph nodes and spleen from rabbits immunized with BSA (Ortiz-Muniz and Sigel, 1967). Antibody peaks in the culture medium of the lymph nodes occurred on the 12th, 42nd, and 48th days of cultivation. Even as late as the 108th day, however, antibodies could be detected in the medium. Primarily 7S antibodies were synthesized.

The antibody titer in the splenic cultures reached a maximum by the 12th day and remained at the same level until the 42nd day, after which the intensity of antibody formation fell, although antibodies could still be detected in the medium on the 84th day.

Basophilic cells with the morphology of lymphoblasts migrated from the explants. They continued to synthesize antibodies even when transferred into a new culture.

The use of homologous serum was shown to be essential for prolonged antibody synthesis. Antibody formation on protein-free medium is weaker and shorter in duration.

Primary Antibody Synthesis In Vitro

Great difficulties have been encountered in the reproduction of the primary immune response *in vitro*, which requires passage through not only the productive, but also the inductive, phase of immunity. Such difficulties were found, for example, in the work of Stevens and McKenna (1957*a,b*) and McKenna (1961).

These workers injected *Salmonella* endotoxin intravenously into rabbits 24 hr before removing their spleens. Fragments of the spleen were then incubated for 1 hr with antigen (bovine γ-globulin). The titers of antibodies that these workers determined by the hemagglutination method were recorded on the 1st, 2nd, and 3rd days of cultivation. Stavitsky (1961) showed that the spleen cells in the experiments of Stevens and McKenna secrete a nonspecific protein material that binds with the antigen, and it is not clear whether this substance is synthesized *de novo* under the experimental conditions or whether it is liberated from the cells.

The first positive results for the induction of a primary immune response were obtained by Fishman (1959, 1961). He based his investigations on the fact that the primary response is composed of two successive phases: the antigen is first phagocytosed and digested by macrophages, after which the processed material is transferred from the macrophages to the precursor cells. In squash preparations from the lymph nodes of immune animals, Thiery (1960) observed groups of lymphocytes surrounding macrophages. It was not clear, however, whether direct cell contact is necessary for the immunological information to be transmitted from macrophages to lymphocytes.

Fishman incubated rat and rabbit peritoneal macrophages *in vitro* with antigen (bacteriophage T2) for 30 min. A homogenate prepared from the macrophages was then added to cultures of lymphocytes in petri dishes on synthetic medium with insulin.

The culture medium was tested for neutralization of phage T2 activity. The results showed that specific antibodies appeared on the 5th day, and could still be found on the 8th and 12th days. On the 5th–7th day of cultivation, some cells had a basophilic cytoplasm, but no plasma cells could be found among them. It is interesting to note that only the homologous test gave positive results; i.e., to obtain a primary immune response, macrophages and lymphocytes from an animal of the same species must be used.

Treatment of the extract of homologous macrophages with ribonuclease abolished the effect. It was accordingly concluded that macrophages produce a special RNA or antigen—RNA complex that is transmitted to the precursor cells and is responsible for inducing antibody synthesis. For this process to take place, direct contact between the cells is unecessary, for cell-free material can cause this phenomenon to develop *in vitro*.

It was later shown that during incubation of peritoneal exudate cells with phage T2, antigens of the tail and head parts of the phage remain in the macrophages (Friedman *et al.*, 1965). The RNA of the macrophages probably acts as an adjuvant, facilitating immunization of the lymphoid precursors by these antigens.

Several investigations have been carried out in the attempt to induce a primary immune response by the use of organ cultures of lymphoid tissue (Auerbach and Globerson, 1965; Luriya *et al.*, 1966*b*, 1967; Tao and Uhr, 1966; Saunders and King, 1966; Holtermann and Nordin, 1968; McArthur *et al.*, 1969). Since lymphocytes survive and proliferate for long periods in organ culture, and since specialized structures can be formed *de novo*, this method could presumably allow the inductive phase of immunity to be reproduced *in vitro*.

The first positive results were obtained by Auerbach and Globerson (1965), who used organ cultures of spleen fragments on millipore filters. An additional essential condition for production of the primary response in these experiments was injection of PHA into the donors 24 hr before removal of the spleen. Without injection of PHA, no positive results could be obtained. The medium consisted of 10% horse serum, 85% Eagle's medium, and 5% embryonic extract. To it was added antigen—a suspension of SRBC. One week later, the cultures were transferred to a new filter to which SRBC were added, and the degree of their agglutination around the fragment was estimated. Investigation of the culture medium during the 2nd week of growth showed that specific antibodies were present in titers of 1:16. Later, these same workers obtained an immune response without the additional injection of PHA (Globerson and Auerbach, 1966). Still later, an antibody response in organ cultures was induced by a number of antigens.

To obtain a primary response, Tao and Uhr (1966) immunized fragments of rabbit lymph nodes with bacteriophage ϕX174, which was added to the culture medium for 2 days starting on the day the cultures were set up. The antibody content in the medium was determined by the phage neutralization reaction. Antibodies were found on the 4th day.

Saunders and King (1966) immunized mixed organ cultures of thymus and spleen fragments from mice with coliphage R17. Antibody synthesis was observed 54 hr after immunization.

These workers rightly point out that with the models they used, further evidence is necessary that the antibody synthesis induced was in fact primary and not secondary, for natural antibodies against coliphage have been found in the sera of some animals.

The author and her co-workers (Luriya *et al.*, 1966*b*, 1967) induced a primary response to horse γ-globulin in organ cultures of whole rabbit and guinea pig lymph nodes. The method of multiple organ cultures in medium 199 with the addition of 10% homologous serum was used. C^{14}-Glycine was added to the medium 24–48 hr before the end of the experiment. After incubation with radioactive label, a homogenate was prepared from the lymph nodes and incubation medium, and antibodies were determined in the supernatant obtained from it by the method of Gurvich and Sidorova (1964), using cellulose–protein copolymers.

The cultures were grown for 10–30 days. Immunization was carried out once on the 4th or 7th day, or three times: on the 1st, 6th, and 24th days of cultivation. Antibody synthesis was found on the 4th and 10th days after immunization *in vitro*.

The immunized cultures were morphologically indistinguishable from the unimmunized. As usual, destruction of most of the lymphoid tissue with preservation of the stroma occurred during the first 4 days. This was followed by regeneration, with the formation of lymphoid follicles in the cortical layer. Regeneration of the medulla was slight, and plasma cells were observed just as infrequently in the immunized as in the unimmunized cultures.

Judging by the increase in radioactivity in the specific immunosorbent compared with the nonspecific immunosorbent, antibody formation in the cultures was significant but of low intensity.

The low level of antibody synthesis in these cultures correlates with the very few plasma cells formed in them. This low level may be attributed to poor regeneration of the medulla, in which most of the plasma cells in the normal lymph node are located. The cortex of lymph nodes in organ cultures regenerates well, however, and contains many living, proliferating lymphocytes.

A primary response *in vitro* has been obtained in organ cultures of fragments of rabbit lymph nodes and spleen incubated for 2 hr with hemocyanin (McArthur *et al.*, 1969). After incubation, the antigen was washed off and the fragments of the lymphoid organs placed on a disk of 0.5% agar surrounded by nutrient medium containing Eagle's medium, vitamins, glutamine, 10% rabbit serum, and C^{14}-glycine (1 μCi/ml).

Antibodies were detected in the culture medium by the passive hemagglutination test, by radioimmunoeletrophoresis, and by the specific coprecipitation test. Specific antibodies appeared on the 3rd day, and were demonstrated until the 20th day. Antibody formation reached its peak between the 6th and 12th days of cultivation (McArthur *et al.*, 1969). Thus, in the last two investigations (Luriya *et al.*, 1966*b*; McArthur *et al.*, 1969), a reliable primary response to antigens introduced for the first time into lymphoid tissue was obtained.

In both cases, antibodies were produced in low titers. Induction of a primary immune response has also been achieved in a suspension of dissociated cells in suspension cultures (Marbrook, 1967; Mosier, 1967; Diener and Armstrong, 1967; Mishell and Dutton, 1966, 1967; Dutton and Mischell, 1967).

In one variant of the experiments, a suspension of lymphocytes in a minimal volume of medium was placed on a cellophane membrane; beneath the cellophane was a reservoir containing medium. The vessel was shaken constantly, the nutrient medium was renewed, and metabolic products were removed. During cultivation of spleen cells of an unimmunized mouse on a dialysis membrane in this way, Marbrook (1967) observed a primary immune response to SRBC and horse red blood cells (HRBC) added to the suspension of lymphocytes. The number of nucleated spleen cells suspended in 1 ml medium was 2×10^7. The nutrient medium consisted of Eagle's medium with the addition of pyruvate and 10% fetal calf serum. Antibody-producing cells appeared on the 3rd—5th day.

Is the immune response to SRBC in fact a primary immune response? This question was under active discussion, since cells capable of producing spontaneous hemolysis of SRBC have been found in the spleens of unimmunized mice. Playfair and co-workers (1965) found 5×10^3 such cells, i.e., 50 cells to every 10^6 nucleated spleen cells. Even so, some workers believe that the immune response in the case of immunization with SRBC is of the primary type, as shown by the predominant synthesis of 19S antibodies and the inhibitory effect of actinomycin D on antibody formation (Bussard and Lurie, 1967; Robinson *et al.*, 1967).

Mishell and Dutton (1966, 1967) grew cultures of spleen cells in plastic dishes rotated eccentrically in the horizontal position. The nutrient medium included Eagle's medium, 10% fetal calf serum, and other components that they describe as a "nutritional cocktail." The dish contained 2×10^7 spleen cells that were immunized with SRBC or HRBC. The cells were tested for antibody formation by Jerne's method, and the culture medium was tested for hemolysins. The number of plaque-forming cells reached a maximum on the 3rd or 4th day of cultivation. In one million nucleated cells there were 10^3 capable of plaque formation; i.e., their number was 3 orders of magnitude greater than initially.

In the control cultures to which no antigen was added, however, the number of antibody-synthesizing cells was 100 times greater (2 orders of magnitude). Perhaps the fetal calf serum contains an antigen that gives a cross-reaction with the red cell antigens. This hypothesis was confirmed by immunoelectrophoretic analysis. Hemolytic antibodies were found in the culture medium of the immunized cells in titers of 1:64 and 1:16, but in the control group, none was produced. Mishell and Dutton (1967) consider that a low oxygen concentration, slight agitation of the suspension, the presence of fetal calf serum, a correct cell density, and daily feeding of the cultures are essential conditions for the induction of the primary response. Under these conditions, an immune response can be produced *in vitro* that is not much weaker than *in vivo*. It must be emphasized that the primary response was induced in the cultures only in a thick suspension of dissociated cells. Although the natural architectonics of the spleen was disturbed in these cases, the high cell concentration and rotation of the cultures allowed adequate interaction among the cells in the system.

That this is so is shown, in particular, by the fact that the production of a primary response in suspension cultures required the presence of two spleen cell fractions: macrophages and lymphocytes (Mosier, 1967). Separation into fractions rich in macrophages and lymphocytes was carried out on the basis of adhesion of the cells to the plastic dishes, and the method of Mishell and Dutton was used for cultivation and immunization. Mosier considers that under these conditions, macrophages process the antigen and transmit it to the lymphocytes. Microfilming experiments confirmed that during plaque formation, contact takes place among particular types of cells, and 30% of plaques were formed around two associated cells (Bussard and Lurie, 1967).

That macrophages are essential for the primary immune response to take place *in vitro* is shown by the results of fractionation of spleen cell suspensions by repeated filtration through a column containing cotton wool (Theis and Thorbecke, 1970). As a result of this procedure, the fraction of "adherent" cells was removed from the cell suspension, and this removal correlated with reduced ability to give an immune response *in vitro* to addition of corpuscular antigens: SRBC or *Brucella abortus* cells. It is interesting to note that macrophages are essential not only for the primary immune response to these antigens, but also for the secondary response (Theis and Thorbecke, 1970).

The fraction of cells that adhere readily to plastic, glass, or cotton wool consists chiefly of macrophages. If these cells are incubated *in vitro* with antigen (SRBC) and then washed off and mixed with the fraction of nonadherent cells, the primary response to SRBC can be reproduced (Pierce, 1969).

When mixed with nonadherent cells, adherent cells (A cells) have a synergic (not additive) effect on the induction of the immune response *in vitro*. The stimulating effect of the A cells on lymphocytes was found to be radioresistant (D_0 = 500–2500 rad), from which it follows that proliferation of the A cells is not required for stimulation. Effectively acting A cells can be obtained from the lymphoid tissue of normal and immune donors. This findings shows that no significant changes in the stimulant activity of the A cells take place during immunization. The chief cell type among the A cells is the macrophages, which actively phagocytose the antigen. Is phagocytosis of the antigen a specific function of the A cells, and what is its role in their interaction with the nonadherent cells during induction of the immune response?

The use of inhibitors of phagocytosis (sodium fluoride, dinitrophenol, iodoacetate) has provided the answer to this question. Treatment with these inhibitors did not abolish the ability of the A cells to involve lymphocytes in the immune response to SRBC *in vitro* (Winfield, 1971). The main role in induction of the immune response is played by the antigen, which has not been phago- cytosed but which is bound to the cell surfaces of the macrophages (Schmidtke and Unanue, 1971; Leserman *et al.,* 1972). Mouse macrophages that were preincubated with human albumin and then treated with trypsin lost their ability to induce the immune response when transferred to a syngenetic recipi- ent. Treatment of A cells preincubated with SRBC and washed off with 0.83% NH_4Cl solution completely prevented the development of the immune response in a population of nonadherent cells. NH_4Cl produces lysis of red cells bound to the surfaces of the macrophages without damaging the A cells themselves, for if the latter are again bound with antigen, they can stimulate immunogenesis in the fraction of nonadherent cells.

The time of contact between adherent and nonadherent cells necessary for induction of the primary response *in vitro* is 2 days (Cosenza and Leserman, 1972). This time coincides with the period during which antigen and T cells must be present.

As Mosier and Coppleson (1968) showed, the cells responsible for synergism of the combined populations occur in the fraction of adherent cells in a concentration of 1:1000–10,000. These active cells of the adherent fraction have been called *cells of the third type.* During immunogenesis *in vitro,* inter- action thus takes place not only among cells of the two types of T and B lymphocytes, but also among cells of a third type, contained in the fraction of A cells. It is assumed that the cells of the third type make contact initially with the antigen, "focusing" it and thus creating optimal conditions for interaction of the T and B lymphocytes.

Cells of immunized cultures that give plaques have been shown to arise from cells that proliferate by mitosis *in vitro,* for which the S-phase commences 24 hr after the beginning of cultivation (Dutton and Mishell, 1967). The addition of thymidine-H^3 in a high concentration (sufficient for "thymidine suicide") leads to selective death of cells that synthesize DNA. If the isotope is added during the first 24 hr of cultivation, the number of plaque-forming cells remains unchanged (does not fall). The addition of thymidine between 24 and 72 hr after the beginning of immunization sharply reduces the number of plaques. Exposure for 8 hr (from 32 to 40 hr) is sufficient for plaque formation to be inhibited completely. All antibody-forming cells thus arise as a result of division of precursors, starting 24 hr after immunization.

By the use of *in vitro* cultures and the method of cell fractionation in a viscosity gradient, Haskill and Marbrook (1972*a,b*) isolated fractions containing most precursors of the plaque-forming cells from a population of spleen cells of normal mice.

They found that at least three discrete populations of precursors of plaque-forming cells exist in the spleen (fractions 2, 2b, and 3). These fractions differ in the kinetics of their response, their antigen requirements, their interaction with radioresistant accessory cells, their presence in the circulation, and their metabolic state (cell cycle). They differ from the precursors of plaque-forming cells of the bone marrow, and they are probably stages in the differentiation of the B lymphocytes. The precursors contained in fraction 2 are sensitive to thymidine-H^3. Among the precursors of the plaque-forming cells of fractions 2b and 3, no thymidine-sensitive cells were found, from which it follows that they do not proliferate. They start to proliferate a short time after the beginning of cultivation with the antigen.

The lag-phase before the appearance of clones of plaque-forming cells *in vitro* differs for different populations: the duration of the lag-phase is 18 hr for fraction 2, 30 hr for fraction 2b, and 44 hr for fraction 3.

The development of a clone of plasma cells *in vitro* is preceded by a stage of blast-transformation, which is associated with an increase in volume of the cell and with a change in its ability to bind antigen.

Cells of the peritoneal exudate, constituting a mixed population of macrophages and lymphocytes, are also capable of giving an immune response to SRBC *in vitro* (Benndinelli and Wedderburn, 1967).

During short-term cultivation of mouse peritoneal exudate cells in medium containing carboxymethylcellulose, SRBC, and complement, zones of hemolysis can be seen to be formed. In the course of 24 hr, about 3000 plaques of hemolysis per 10^6 explanted cells appear, the first plaques being visible 15 min after the beginning of incubation *in vitro*. Hemolysis of the red cells in this system has been shown to take place by the action of specific antibodies that are synthesized *de novo* (Bussard *et al.*, 1970).

Addition of actinomycin blocks antibody synthesis, which consequently depends on the synthesis of m-RNA. The kinetics of hemolytic activity of the exudate cells indicates that the increase in the number of plaque-forming cells does not take place through proliferation, although it is due to the appearance of new plaque-forming elements (Bussard *et al.*, 1970). It is interesting to note that cells of the lymph nodes, thymus, and spleen do not form plaques of hemolysis under these conditions, and if these cell suspensions are cultivated together with exudate cells, the number of plaque-forming cells in the latter is reduced. This phenomenon is still largely a matter for conjecture. It is probably a specific immune response. However, it is difficult to conclude what type of course (primary or secondary) it pursues.

Besides erythrocyte antigens, other antigens were also successfully used for induction of antibody responses in suspension cultures, the flagellin of *Salmonella adelaide* being one of them. The intensity of this reaction was assessed by the method of bacterial immunocytoadhesion. The number of antibody-forming cells reached a maximum on the 4th day after immunization. Its level was 280 antibody-synthesizing cells per 10^6 surviving cells, falling on the 6th day to 78 per 10^6 cells (Diener and Armstrong, 1967).

With a low concentration of explanted cells (5×10^6 per ml), no immune response is induced, probably because the cell contacts that provide the necessary intercellular interaction for immunization do not take place under these conditions.

2. The Delayed Hypersensitivity Reaction *In Vitro*

Basic Mechanisms of the Delayed Hypersensitivity Reaction In Vivo and Aims of Research into This Process In Vitro

The reaction of lymphoid tissue to foreign antigens is shown not only by antibody formation, but also by the development of delayed hypersensitivity. This type of immune response is most marked after the transplantation of foreign tissue and intradermal injection of antigens. The cells directly responsible for delayed hyposensitivity reactions (DHRs) are T lymphocytes. They infiltrate the graft or area of skin containing the foreign antigen and induce tissue destruction. Humoral antibodies do not play an essential role in this process. It is important to note that several autoimmune diseases (Hashimoto's disease, acute glomerulonephritis, rheumatic polyarthritis, etc.) follow a course of the DHR type. In these cases, the immunological activity of the lymphoid tissue is directed against tissue-specific antigens of the host organism itself.

The morphology of the DHR has been well studied in connection with allograft rejection. A series of successive morphological changes is observed in the regional lymph node. During the 1st day, vacuoles surrounded by a single-layered membrane appear in the small lymphocytes of the regional lymph node. After 48 hr, the vacuolated cells disappear. On the 4th and 5th day, many pyroninophilic lymphoid cells accumulate in the cortical zone of the lymph node; they measure 11–20 μm in diameter, and they have one or more nucleoli and a vesicular nucleus. The cytoplasm of these cells is strongly basoplilic, it contains many free ribosomes, and as a rule it possesses no ergastoplasm. The number of these cells increases until the 11th day. They are not connected with secondary follicles, but lie in the cortical region around them. The accumulation of large pyroninophilic cells in the lymph nodes coincides with the period of resorption of the allograft.

Rejection of the graft takes place through small lymphocytes with hypertrophied cytoplasm, and of histiocytes that infiltrate the graft. The infiltration process begins on the 5th–8th day. Destruction of the foreign tissue takes place during close contact with lymphocytes, and many of the lymphocytes themselves die. The DHR is a T-cell response. It requires interactions among several subclasses of T cells (including helper cells, killer cells, and suppressor cells) and between T cells and macrophages (Bullock *et al.*, 1975; Rosenthal and Shevach, 1973).

Many aspects of the transplantation immunity and delayed hypersensitivity reactions have not yet been studied. For instance, the cellular mechanisms of identification of foreign antigens during induction of the DHR are still unknown, it is not clear whether direct cell contact between the lymphocytes and the tissue is necessary for rejection, or by what mechanism the allogeneic graft is destroyed; differences in the relationship between the immune and nonimmune lymphocyte and the target cell await explanation. Most T-cell-mediated immune functions have been shown to depend on structures coded in the major histocompatibility locus, and the question arises whether H-2 compatibility between T cells and the cells with which they cooperate in immune reactions is always required (Shreffler and David, 1975).

To answer these and other questions, the *in vitro* models of the DHR and of its individual components have been widely used. The great practical importance of development of *in vitro* models of the DHR by which the degree of immunological compatibility between potential donors and the recipient can be assessed must be borne in mind. The creation of adequate systems in tissue culture could also help to solve some problems connected with the diagnosis of autoimmune diseases.

Blast-Transformation, Recognition of Antigens, and the Cellular Immune Response In Vitro

Blast-transformation of lymphocytes takes place in suspension cultures of peripheral blood leukocytes from two individuals or two nonsyngenetic animals, as a result of which large basophilic cells similar to those observed in cultures stimulated by PHA appear (Bain *et al.*, 1964; McLaurin, 1965; Elves *et al.*, 1966; Schwarz, 1966). In contrast to PHA-stimulated cultures, in which blast cells account for up to 95% of all lymphocytes and the intensity of the reaction reaches its maximum on the 2nd or 3rd day, in mixed cultures of lymphocytes, only 10–20% undergo transformation, and the reaction reaches maximal intensity after 6 or 7 days (Caron and Sarkany, 1966). In cultures of lymphocytes from identical twins, transformation does not occur (Bain *et al.*, 1964; Chalmers *et al.*, 1966).

If the lymphocytes of F_1 hybrids are mixed in culture with the lymphocytes of one of the parental strains, only the lymphocytes of that parental strain that become immunized with antigens of the other parental strain undergo blast-transformation, and the lymphocytes of the F_1 hybrid themselves are not transformed. More intensive blast-transformation can be obtained by preliminary immunization of one of a pair of allogeneic partners with the isoantigens of the other (Rubin *et al.*, 1964; Oppenheimer *et al.*, 1965). If lymphocytes from mutually tolerant donors are grown together in culture, however, no blast-transformation reaction is observed *in vitro*, and the culture morphology is similar to that of unstimulated cultures from one donor. These observations, together with the fact that administration of puromycin completely inhibits blast-transformation in mixed cultures, demonstrate that blast-transformation of lymphocytes is a manifestation of a primary immune response to transplantation antigens in tissue culture. This reaction does not lead to destruction of the lymphocytes, but it stops at the stage of formation of cells with immunoblast morphology. In other words, the recognition of foreign transplantation antigens in mixed cultures of lymphocytes is manifested as blast-transformation.

The reaction of lymphocytes in mixed cultures of white blood cells from two donors can be restricted to one direction. For this purpose, lymphocytes from one donor are treated with mitomycin C, as a result of which the cells do not lose their antigenic properties, but do lose their ability to undergo blast-transformation and proliferation (Adler *et al.*, 1970). The intensity of the reaction in such cultures can be estimated from thymidine-H^3 incorporation into the stimulated cells, autoradiographically, and by counting the blast cells. It has been shown that the conditions of cultivation have an important effect on the intensity of proliferation of lymphocytes in response to immunization. If lymphocytes are grown in culture above a dialysis membrane together with cells treated with mitomycin C in a tube lowered into a vessel containing nutrient medium, as described by Marbrook, incorporation of radioactive thymidine into the lymphocytes is 34 times greater than if the cultures are grown in ordinary tubes (Wagner and Feldmann, 1972). With Marbrook's method, the period of growth of lymphocyte cultures can be prolonged (for 6 or 7 days), the time necessary for the maximal immune response to tissue compatibility antigens to occur. Lymphocytes immunized under these conditions exhibit specific cytotoxic activity toward target cells, which can be recorded as liberation of radioactive chromium from prelabeled targets. Interaction between lymphocytes immunized *in vitro* and target cells takes place in a petri dish seeded with 5×10^4 target cells to which 5×10^6 living lymphocytes have been added. The rates of lysis of the target cells *in vivo* and *in vitro* are similar. The lag-phase lasts 30–50 min, and 100% lysis occurs after 200 min (Wagner and Feldmann, 1972). The reaction of cellular immunity, which consists of a phase of recognition of the antigen followed by an effector phase, can thus be reproduced completely in tissue culture.

Reproduction of cellular immunity *in vitro* has yielded much important information on the mechanism of this reaction, which remained unexplained after the use of *in vivo* systems. In particular, this applied to the role of macrophages and B cells in the induction of the immune response to tissue compatibility antigens. In a mixed lymphocyte culture (MLC), the macrophages perform the nonspecific function of assistance: macrophages from immune and nonimmune donors are equally effective when added to the cultures (Rode and Gordon, 1970).

Cell transfer experiments *in vivo* carried out with the lymphoid tissue of neonatally thymectomized animals and also with separate populations of T and B lymphocytes (J. F. A. Miller *et al.*, 1971) have shown that the T lymphocytes play the leading role in the mechanism of cellular immunity reactions. This finding, however, does not rule out completely the possible participation of B lymphocytes in cellular immunity. Some light has been shed on this problem by the use of *in vitro* systems and fractionation of the cells by preparative electrophoresis, giving pure populations of T and B lymphocytes, which differ in their electrophoretic mobility. The cell fraction with high mobility is a pure population of T cells; that with low mobility, a pure population of B lymphocytes (Raff and Wortis, 1970; Häyry *et al.*, 1972). Lymphocytes of the high-mobility fraction reacted to foreign antigen in mitomycin-treated MLC, whereas lymphocytes of the fraction with low mobility did not possess this property. Mixing the cells of these two fractions did not given an enhanced response. These observations show that it is unnecessary for T lymphocytes to cooperate with B-cells in order to react to allogeneic antigens (Häyry *et al.*, 1972).

Active cell division is necessary during development of the immune response *in vitro*. Elimination of proliferating cells leads to complete abolition of the cytotoxic effect. Investigations of cell fractionation in a viscosity gradient by the method of R. S. Miller and Phillips (1969) enabled a morphological analysis to be made of the population of effector cells (Häyry *et al.*, 1972). Fractionation of cells from a 6-day unidirectional MLC showed that most cells responsible for lysis of the target cells are contained in the blast fraction, which includes cells over 9 μm in diameter. When the blast fraction was grown in culture for a further 4 days, the cells were reduced in size, but they retained their immunological activity. Hence, it follows that the morphology of the "killer" cells differs at different phases after immunization. In the early stage, they are large blast cells, which later decrease in size while retaining their cytotoxic activity.

Besides lymphocytes, macrophages can also act as effector cells. Cooperation between lymphocytes and macrophages is not essential in the effector phase. The mechanism of action of macrophages that behave as effector cells *in vitro* has not been completely elucidated. The suggestion has been made that macrophages act as effectors under the influence of a soluble mediator from the lymphocyte (Evans and Alexander, 1970). Whether this mediator consists of antibodies cytophilic for macrophages has yet to be clarified.

A primary transplantation immunity reaction has also been reproduced in full in another system *in vitro* (Ginsburg, 1968).

Rat lymphocytes were grown on a feeder monolayer of mouse embryonic cells. Large pyroninophilic cells that caused lysis of the subjacent mouse cells appeared under these conditions. The immune response thus produced showed strict specificity, which was demonstrated by transferring the immune lymphocyte population into other cultures, where embryonic cells from mice of strains differing from the first in their H-2 tissue compatibility antigens were grown.

When immune lymphocytes are explanted, they adhere firmly in the course of 2 hr to fibroblasts of the embryos that were used for immunization *in vitro* (Lonai *et al.*, 1972). The adherent cells are responsible for the cytotoxicity reaction against fibroblasts of that particular genotype. After 48 hr, some of the adherent cells pass into the supernatant. On repeated incubation of lymphocytes from the supernatant on fibroblasts, complete elimination of lymphocytes reacting to that particular genotype can be obtained.

In Vitro Destruction of Target Cells by Lymphocytes from Preimmunized Donors

If lymphocytes from a preimmunized donor are grown in monolayer culture together with living allogenetic target cells, the latter are destroyed (Govaerts, 1960; Merrill *et al.*, 1960; Rosenau and Moon, 1961; Brondz, 1968). Lymphocytes obtained from various sources—the thoracic duct, lymph nodes, spleen, peritoneal exudate, and peripheral blood—have a cytotoxic effect. This cytotoxic action is characterized by strict immunological specificity; the lymphocytes have a cytotoxic effect on the cells of those animals the antigens of which were used to sensitize the donor (Rosenau and Moon, 1961; D. W. Wilson, 1963; Holm, 1966). The presence of humoral antibodies and complement is not essential for the cytotoxic effect, but direct contact between lymphocytes and target cells is necessary. Separation of the immune lymphocytes by a millipore filter prevents lysis of the target cells.

The morphology of the cytotoxic effect has been studied in detail (Rosenau and Moon, 1962; Brondz, 1968; D. W. Wilson, 1965a,b; Möller, 1965). In Rosenau and Moon's experiments, lymphocytes from immunized BALB/C mice were added to cultures of fibroblasts from C3H mice growing in Leighton's tubes. After 18 hr, the lymphocytes were brought into contact with the fibroblasts, as a result of which the latter became rounded, lost their processes, and developed vacuoles. Most of the cells were destroyed between 18 and 40 hr, fibroblasts as well as lymphocytes. It was later shown that lymphocytes from immunized donors have a cytotoxic effect on monolayer cultures of allogenetic tumors, macrophages, and epithelial cells. Determination of the time required for manifestation of the cytotoxic effect yielded interesting results. When lymphocytes from regional lymph nodes of immune rats were added to monolayer cultures of target cells from an allogenetic tumor, D. W. Wilson (1965a) observed adherence of the immune lymphocytes to the surfaces of the target cells after 7 hr, while after 24 hr or more, destructive changes were found in the target cells; these changes reached a maximum after 48 hr.

The number of target cells destroyed *in vitro* depends on the number of sensitized lymphocytes added to the culture (Brunner *et al.,* 1966; D. W. Wilson, 1965*a,* 1967).

In monolayer cultures of low density, target cells are destroyed in about 6 hr (Granger and Weiser, 1964).

If the ratio of lymphocytes to target cells is 100:1, lysis takes place in 2 hr. By a microfilming technique, it was found that one lymphocyte can react with several target cells, lysing them, yet itself remaining viable. Active protein synthesis in the lymphocytes is essential for the cytotoxic effect. That it is is shown by suppression of the cytotoxic effect by the addition of hexamine and actinomycin D.

The cytotoxic effect *in vitro* can also be inhibited by the addition of 6-mercaptopurine (D. W. Wilson, 1965*b*) or hydrocortisone (Rosenau and Moon, 1962, 1966), and by treatment of the lymphocytes with ultrasound, heat, or freezing and thawing.

Addition of immune serum to the culture medium depresses or completely abolishes the cytotoxic effect of lymphocytes on target cells; on the other hand, the serum from unimmunized animals has no effect on this process (Möller, 1965). Inhibition of the cytotoxic effect through the action of the immune serum is probably attributable to the fact that humoral antibodies bind to the antigenic determinants of the target cells, blocking them and preventing interaction with the immunocompetent cells (Möller, 1965; D. W. Wilson, 1965*b*).

To explain the mechanism of interaction of the immune lymphocytes with the target cells, it was postulated that the immune lymphocyte carries on its surface the determinant of the antibody responsible for specific attachment of the lymphocyte to the cell with the corresponding antigen. This hypothesis has been confirmed experimentally and its details clarified by a series of investigations (Govaerts, 1960; Rosenau and Moon, 1961; Brondz, 1968).

Lymphoid cells from immunized donors proliferate in culture when antigens for which they are specific are added to the medium. This proliferative response could be determined by either pulsing the cells with radiolabeled DNA precursor or estimating the number of blasts in cell smears. In humans, the lymphoid cells were usually obtained from peripheral blood, and the antigen most frequently used was purified protein derivative of tuberculin. In rabbits, guinea pigs, and mice, cells were obtained from spleen and lymph nodes, and the antigens included soluble proteins such as albumins, Ig, tetanus, and diphtheria toxoid. The presence of adherent cells or purified macrophages appears to be important in order for preimmunized lymphocytes to respond to antigens *in vitro* (Hersh and Harris, 1968; Unanue, 1972). The response required close lymphocyte–

macrophage contact. Thus, separation of the macrophages from the lymphocytes by a millipore filter (pore size 0.45 μm) impaired the response. It was found by Rosenthal and Shevach (1973) that in the response to soluble protein antigens, the cooperation between macrophages and lymphocytes of guinea pigs required identity at the major histocompatibility locus.

This antigen recognition by primed lymphocytes can be used as an *in vitro* analogue model of the initial steps of delayed hypersensitivity reactions.

The cytotoxic effect in culture can be reproduced by using lymphocytes from a nonimmune donor if special conditions are observed (Perlmann *et al.*, 1968; Holm and Perlmann, 1967). If cells in suspension or in a monolayer are treated with antigen (thyroglobulin or serum albumin) that is absorbed on their surface, and if a specific antiserum is then added to the incubation medium, an antigen–antibody complex is formed on the surfaces of the cells. If nonimmune syngenetic, allogenetic, or xenogenetic lymphocytes are added to this system, they adhere to the cells that carry this complex. Lysis of the target cells is observed later. In this case also, direct contact between target cells and lymphocytes is essential for lysis, and the lymphocytes remain viable during the reaction. The reaction is inhibited by antimetabolites and by antilymphocytic sera. It is assumed that for the lymphocytes to be able to act on the treated target cells, the lymphocyte must be activated by the antigen–antibody complex on the surfaces of the target cells.

Substances that stimulate mitosis have a specific action on the population of nonimmune lymphocytes. These substances include PHA (Holm and Perlmann, 1965, 1967, 1969; Möller, 1965; Hellström and Hellström, 1967), streptolysin, a streptococcal filtrate (Holm and Perlmann, 1965), and other agents. Lymphocytes stimulated by these substances not only increase mitotic activity and stimulate blast-transformation, but also exhibit cytotoxic activity when added to target cells. The cytotoxic effect correlates with the level of DNA synthesis and with the blast-transformation reaction of the lymphocytes induced by mitogens. Presumably, changes in the surface properties and mobility of the activated lymphocytes play some part in this reaction.

The cytotoxic effect *in vitro* is a two-stage reaction that includes activation of lymphocytes and lysis of the targets. If the lymphocyte population is obtained from a previously immunized donor, protein synthesis in the lymphocytes is essential for reproduction of the cytotoxic effect in culture. If other models (formation of an antigen–antibody complex on the target cell surface, activation of lymphocytes by mitogens) are used, protein synthesis is not an essential condition for the reproduction of the cytotoxic reaction. In all the systems described above, however, direct contact between lymphocytes and target cells is necessary.

The essence of the matter is that in an experimental model to reproduce destruction of target cells by lymphocytes from a previously immunized donor, most of the lymphocytes added to the culture adhere to target cells, whereas immune cells form only a very small proportion of the lymphoid population. The presence of a small percentage of sensitized cells is thus sufficient to produce adherence of the nonimmune lymphocyte population to the target cells. The mechanism of this phenomenon is unexplained. It may be influenced by certain humoral factors.

Lysis of target cells *in vitro* may take place not only as a result of their direct contact with lymphocytes, but also through the action of cytotoxic substances produced under certain conditions of cultivation. For instance, in cultures of lymphocytes obtained from donors in which the DHR (e.g., to tuberculin, bovine γ-globulin, egg albumin, etc.) has first been developed in the presence of this antigen, a cytotoxic substance that destroys target cells accumulates in the culture medium, although the target cells are not related to the antigen used for sensitization (Ruddle and Waksman, 1968).

To reproduce the effect, it is not essential to incubate the lymphocytes treated with the antigen and the target cells together. The supernatant obtained by centrifuging the medium in which lymphocytes from a tuberculin-sensitive donor were grown in the presence of tuberculin for 17 hr or more has been shown to possess cytotoxic activity.

It is interesting to note that this lysis effect is not linked to genetic differences between lymphocytes and target cells. By the use of embryonic fibroblasts from inbred rats as targets and lymphocytes from the same donors, lysis of the fibroblasts takes place in 72 hr if the antigen to which the lymphocytes were sensitized is present in the medium.

The phenomenon described above has been reproduced with the use of sensitized lymphocytes from mice, rats, guinea pigs, and man (Granger *et al.*, 1969). In all cases, the lymphocytes, when incubated with the corresponding antigen, liberated into the medium substances that led to destruction of the target cells.

Culture medium containing cytotoxic substances likewise has a stimulant action on lymphocytes that leads to their proliferation (Wolstencroft and Dumonde, 1970). If it is added to the suspensions of syngenetic or allogenetic lymphocytes, whether from immune or nonimmune donors, intensive incorporation of thymidine-H^3 is observed, reflecting increased mitotic activity of the cells. In the presence of antigen, the immune lymphocytes thus secrete a factor that is mitogenic for lymphocytes and that in the opinion of Wolstencroft and Dumonde, may play an important role in the regulation of the immune response *in vivo*.

Secretion of cytotoxic substances and their liberation from the cells can take place not only in cultures of lymphocytes from donors with a DHS in the

presence of antigen, but also during incubation of lymphocytes from nonimmune donors in the presence of several agents that induce blast-transformation
of lymphocytes. These agents include PHA, streptolysin, xenogenetic antibodies,
and antilymphocytic serum. For instance, on the addition of PHA to a culture of
lymphocytes obtained from human tonsils, toxic substances are secreted into
the medium after 2 or 3 hr, i.e., long before the appearance of morphological
changes in the lymphocytes themselves. In the course of cultivation (up to 96
hr), the toxic action of the medium increases. It is not necessary for PHA to be
constantly present in the medium: for production of the toxic factor to take
place, PHA need be present in the medium only 15 min, after which it can be
washed out (Williams and Granger, 1969).

Thus, several models that can be used to study the destruction of target cells
by lymphocytes *in vitro* are now known.

1. Destruction of target cells by the action of lymphocytes sensitized *in
vitro* and *in vivo* by transplantation antigens of the targets. The cytotoxic action
is effected by direct contact between lymphocytes and target cells.

2. Destruction of target cells by cytotoxic substances that are produced by
lymphocytes sensitized to foreign antigens when added to the culture medium.

3. Destruction of target cells by contact interaction with nonimmune
lymphocytes and by the action of cytotoxic substances secreted by these
lymphocytes when treated with PHA and other mitogens.

Characteristically, the second model can reproduce the cytotoxic reaction
against syngenetic target cells.

It is therefore natural to expect that contact between the antigen and a
population of cells isolated from the lymphoid tissue of a sensitized donor and
added to the culture may have a complex effect: under these conditions, some
cells may be stimulated to proliferate, while others may play the role of target
cells subjected to lysis. In this connection, the formation of colonies of fibroblasts in monolayer cultures of lymphoid and hematopoietic cells provides an
interesting model. As has already been mentioned, cells from bone marrow,
spleen, and peritoneal exudate when grown in monolayer cultures form colonies
of fibroblasts that are clones. Each of these cell populations has its own
particular concentration of precursors of the fibroblast colonies. Clones of
fibroblasts are thus formed in systems that contain immunocompetent cells.
Does the presence of antigen in the medium have any effect on this process if
the cells are taken from a sensitized donor? The answer to this question was
obtained by the use of sensitization to tuberculin as the model.

Suspensions of bone marrow, spleen, and peritoneal exudate cells were
obtained from normal guinea pigs and from guinea pigs immunized by intradermal injection of living BCG vaccine or by subcutaneous injection of Freund's
complete adjuvant to which killed dried BCG vaccine was added (Luriya *et al.*,

TABLE 12. Effect of Tuberculin on the Formation of Fibroblast Colonies by Cells of Normal and Sensitized Donors[a]

Donor No.	Condition	Cells explanted	Tuberculin in medium	Number of colonies[b]	Degree of inhibition
1	Normal	Bone marrow $1 \times 10^{5\,c}$	−	26, 29	1
			+	21, 25	
		Bone marrow $5 \times 10^{6\,c}$	−	68	1
			+	70, 74	
2	Normal	Macrophages 2.5×10^5	−	16, 25, 28, 29	1
			+	21, 30, 33	
3	Normal	Macrophages 4.5×10^5	−	72, 85	1
			+	70, 107	
4	Normal	Bone marrow 17×10^6	−	41, 87	1
			+	41, 74	
5	BCG 35 days	Macrophages 10^6	−	173, 188	45
			+	3, 4	
		Bone marrow 20×10^6	−	290, 366	1.3
			+	233, 234	
		Spleen 30×10^6	−	372, 442	1.8
			+	205, 251	
6	BCG 52 days	Macrophages 34×10^5	−	66, 86	4
			+	17, 23	
7	BCG 52 days	Macrophages 34×10^5	−	135, 347	120
			+	0, 3	
8	BCG 75 days	Macrophages 10^7	−	405	130
			+	3	
9	BCG 75 days	Macrophages 18×10^5	−	107, 119	4
			+	29, 32	
10	BCG 75 days	Macrophages 2×10^6	−	39, 45, 111	9
			+	4, 7, 12	
11	BCG 75 days	Spleen 10^7	−	53	0.8
			+	56, 57, 88	

1972a). The cells were grown in Roux flasks in medium 199 with the addition of bovine serum. Tuberculin was added to half the cultures. On the 10th–12th day, the number of colonies of fibroblasts in the cultures was counted.

The presence of tuberculin in the medium had no effect on colony formation in cultures of bone marrow, spleen, and peritoneal exudate of normal donors (Table 12). If no tuberculin was added to the nutrient medium, cultures from sensitized donors had the normal morphology. The addition of tuberculin to the medium, however, led to a substantial change in the effectiveness of colony formation by one category of sensitized cultures. The results presented in Table 12 show that the presence of tuberculin had no effect on the number of colonies in cultures of spleen and bone marrow cells from vaccinated guinea pigs. Bone marrow and spleen cells from only one donor, taken on the 27th day after

TABLE 12 (cont.)

Donor		Cells explanted	Tuberculin in medium	Number of colonies[b]	Degree of inhibition
No.	Condition				
12	BCG 75 days	Bone marrow 27×10⁶	−	96	1
			+	65, 137	
		Spleen 6×10⁷	−	93	0.8
			+	112	
13	Adjuvant 18 days	Macrophages 6.7×10⁵	−	>500	1
			+	>500	
14	Adjuvant 27 days	Spleen 20×10⁶	−	35, 69, 79	1.3
			+	27, 52, 58	
		Macrophages 9×10⁵	−	>500	1
		Bone marrow 26×10⁶	+	>500	
			−	190	1
			+	128, 138	
15	Adjuvant 27 days	Macrophages 3.5×10⁵	−	283, 342	3
			+	78, 153	
		Spleen 34×10⁶	−	159, 353	3
			+	69, 70	
		Bone marrow 40×10⁶	−	174, 184	2
			+	88, 93	
16	Adjuvant 35 days	Macrophages 14×10⁵	−	305, 343, 360	77
			+	3, 5, 6	
		Spleen 34×10⁶	−	468, 566	1
			+	448, 493	
		Bone marrow 26×10⁶	−	154, 166	1.6
			+	93, 115	

[a]After Luriya et al. (1972).
[b]Each number is the number of colonies in one culture vessel.
[c]Cultures grown in plastic dishes.

vaccination with Freund's adjuvant, showed a decrease of half and two-thirds, respectively, in the intensity of colony formation. Meanwhile, the addition of tuberculin to cultures of peritoneal exudate of the vaccinated donors led to a sharp decrease in the intensity of colony formation (Fig. 47). The degree of inhibition of the number of colonies in cultures of exudate from guinea pigs immunized with living BCG vaccine was between 4- and 130-fold; in the case of peritoneal exudate cultures from donors sensitized 35 days before the cell suspensions were obtained by injection of Freund's adjuvant, the degree of inhibition was 77-fold. In some cultures, virtually complete inhibition of colony formation occurred in the presence of tuberculin. At the same time, in cultures of exudates taken on the 18th and 27th days after sensitization with Freund's adjuvant, the presence of tuberculin caused virtually no decrease in the efficiency of colony formation. It is noteworthy that injection of Freund's

Fig. 47. Inhibition of formation of fibroblast colonies in cultures of peritoneal exudate cells from tuberculin-sensitive guinea pigs by tuberculin present in the nutrient medium. Left: control culture; right: culture with tuberculin added to medium.

adjuvant into the donors could influence the increase in the number of colony-forming cells among the peritoneal macrophages. The same sharp decrease in the efficiency of colony formation as in cultures of peritoneal exudate cells was also observed in cultures of lymph node cells from sensitized donors in the presence of tuberculin in the medium. That virtually no decrease in the efficiency of colony formation (ECF) was observed in cultures of the bone marrow and spleen, whereas the ECF in cultures of peritoneal exudate and lymph node cells was sharply reduced, may be attributable to the number of immunocompetent cells in each of these populations in subcutaneous immunization experiments.

The system just described has two features that distinguish it from the other models:

1. The inhibition effect is assessed, not by the survival of target cells, but by their ability to form clones, i.e., by the intensity of their proliferation.

2. The inhibition effect is manifested in the system's own cell population: the target cells are not added to the system from outside, but are natural members of the immunologically active cell population.

Inhibition of Macrophage Migration

Cooperative interaction among cells in immunologically active cell populations is also clearly demonstrated in another model: the phenomenon of inhibition of migration of macrophages.

This model can be used to analyze the role and behavior not only of lymphocytes, but also of macrophages in immunological processes. If lymph nodes, the spleen, and suspensions of peritoneal exudate cells from donors in which the DHR has been induced are explanted *in vitro* in the presence of a specific antigen, inhibition of migration of macrophages is observed (Carpenter, 1963; David *et al.*, 1964). If material from a donor producing only antibodies is used for cultivation, migration of the macrophages is not inhibited in the presence of antigen.

Only a few immune lymphocytes, forming a very small percentage of the population, are sufficient to inhibit migration of the macrophages. The phenomenon also takes place if the peritoneal exudate from a donor showing the DHR is diluted 50 times with peritoneal macrophages from an unimmunized animal. Migration of macrophages in this case is inhibited in the whole population (B. R. Bloom and Bennett, 1966; David, 1966). It can be postulated from these results that migration of the macrophages is inhibited by means of a humoral mediator formed by contact between immune lymphocytes and antigen, and that the inhibition is not the result of direct cellular interaction. Such a possibility is indicated by inhibition of migration of macrophages from unsensitized fragments of lymphoid organs if the fragments are grown in the same culture vessel as fragments from hypersensitive donors in the presence of antigen. In other words, a factor that prevents the migration of macrophages from nonimmune lymph nodes is formed in the nutrient medium of cultures of lymph nodes from guinea pigs hypersensitive to bovine γ-globulin on the addition of antigen (Halpern *et al.*, 1967).

The factor that inhibits migration of the macrophages appears in the medium after contact between lymphocytes and antigen for 6–8 hr, and its production continues for 4 days (Thor *et al.*, 1968; Svejcar *et al.*, 1968). The appearance of the factor that inhibits migration of macrophages in the medium is blocked by inhibitors of protein synthesis. Hence, it follows that the factor is synthesized in the cells *de novo*, and is not simply liberated on contact with the antigen. Several biochemical characteristics of the factor that inhibits migration of macrophages are known. It is not dialyzed, and it withstands heating to 56°C for 30 min. It is unaffected by ribonuclease and deoxyribonuclease, but is destroyed by proteolytic enzymes. The molecular weight of the factor is 60–80,000. Electrophoretic studies have shown that it contains albumin and α-globulin.

The factor is not species-specific. Lymphocytes from the blood of a person with a DHR, when incubated with antigen, liberate a factor that inhibits migration of macrophages not only from explants of human lymph node and exudate, but also from guinea pig exudate.

The mechanism of action of the factor that inhibits migration is not yet known. Presumably, macrophages adsorb the factor on their surfaces, thereby producing a change in some of the properties of these cells (lowering of the surface potential, a change in the permeability of the membrane, and in particular a change in its adhesiveness to glass). These changes are probably responsible for inhibiting migration.

It must be asked whether the factor that inhibits migration of the macrophages does so purely by inhibiting their movement or whether it also brings about other changes, not yet studied, in the properties of these cells. It is likewise not known whether the action of this factor is specific purely for macrophages or whether other cells also respond to it, e.g., by a change in motility and differentiation.

The model of inhibition of migration *in vitro* can help to solve a number of clinical problems. It can be used to study autoimmune diseases, immunity against tumors, and immunity against some infectious diseases.

The models we have examined above reproduce *in vitro* individual stages of the transplantation immunity and delayed hypersensitivity reactions.

The Simonsen Phenomenon

Yet another immunological process—the Simonsen phenomenon—has also been reproduced in culture. This phenomenon consists of splenomegaly in a chick embryo and the formation of small foci on the chorioallantois after injection of nonsyngenetic lymphocytes. It is caused by the immunological activity of the transplanted cells (Simonsen, 1957). Enlargement of the spleen in embryos after transplantation of fragments of spleen from adult hens to the chorioallantois was described by Dantschakoff (1924), who suggested the correct explanation of this phenomenon: enlargement of the recipient's spleen is the result of its colonization by cells from the transplanted tissue.

The Simonsen phenomenon belongs to the same category of phenomena as runt disease in mammals, one manifestation of which is splenomegaly. It occurs, in particular, after nonsyngenetic hematopoietic cells from adult donors are injected into newborn mice. This phenomenon in mammals has been reproduced in organ cultures of the spleens of newborn F_1 hybrids on the addition of lymphocytes from an adult donor of the parental strain (Auerbach and Globerson, 1966; Globerson and Auerbach, 1967). Intensive hypertrophy of the explanted spleen is observed in these cultures; addition of syngenetic lymphocytes or of lymphocytes from an irradiated nonsyngenetic donor does not produce splenomegaly.

Spleens of mouse embryos or fragments of the spleens of newborn mice were grown on a millipore filter in Eagle's medium with the addition of horse serum and chick embryo extract. In the case of combined cultivation with syngenetic lymphocytes, no increase in the size of the explant was observed. On the other hand, on the addition of lymphocytes from a donor of the parental line, an appreciable increase in the size of the explanted spleen was found in most cases.

Histological study of the enlarged spleens showed that on the 3rd day, they contain numerous lymphoid cells scattered throughout the explant and not organized into structures. On the 4th day, the number of lymphocytes was reduced, and at the same time, intensive granulopoiesis was developing in the explants. If cultivation continued, macrophages and histiocytes became the predominant cell types.

Using splenomegaly as the model (in the graft vs. host system *in vitro*), the degree of immunological competence of embryonic hematopoietic cells from various sources was estimated (Umiel *et al.*, 1968). Embryonic liver cells do not induce splenomegaly in culture if obtained directly from the embryo or if previously grown in organ cultures for 4–6 days. However, they can acquire the property of inducing splenomegaly if grown in combined culture with thymus tissue.

Immunological Tolerance

Immunological tolerance to polymerized flagellin from *Salmonella adelaide* has been induced in cultures of normal mouse spleen cells (Diener and Feldman, 1972). Induction of tolerance *in vitro* took place in two stages: (1) adhesion of antigen to the surface of the immunocompetent cell in a concentration sufficient to induce tolerance; and (2) inactivation of the immunocompetent cell by antigen bound to the cell surface.

Treatment of lymphocytes with trypsin solution during the 16-hr period of induction of tolerance *in vitro* led to abolition of tolerance. If the trypsin treatment was given 3 days later, however, reversion was not observed. Treatment of spleen cells with trypsin before exposure to the antigen prevents the induction of tolerance *in vitro*.

The use of tissue cultures has led to unexpected results in the analysis of radiation chimeras and of mice in which tolerance was induced by the injection of allogenetic cells during the period of late embryogenesis (I. Hellström and K. E. Hellström, 1973). In all these cases, the lymphocytes behaved *in vitro* not just like tolerant cells, but rather as immune cells, although a state of tolerance was present *in vivo*. If, in fact, fibroblasts of radiation chimeras or of strains to which tolerance was induced were used as the target cells, Hellström found that they underwent lysis if lymphocytes of radiation chimeras or of tolerant donors were added to their cultures. This effect was abolished by the addition of sera of radiation chimeras or tolerant animals to the cultures. It is important to note that similar results were obtained by I. Hellström *et al.* (1969) with tumor target cells when lymphocytes from tumor-carriers were added to them *in vitro*, and also in allophenic mice. Tetraparental (allophenic) mice were obtained by fusing embryos of two parental lines at the 8-cell stage, as a result of which all tissues of the adult mice constituted a cell mosaic. The problem that arose was: does the lymphoid tissue of these animals exhibit true tolerance toward their own cells? To solve this problem, lymphocytes of various donors were added to cultures of fibroblasts of the parental lines. Fibroblasts of the

142 CHAPTER IV

parental line were found to be more effectively destroyed *in vitro* by lympho-cytes of the tetraparental mice than by lymphocytes of the F_1 hybrids or of either parental strain. The cytotoxic activity of the lymphocytes of the allo-phenic mice was abolished by the addition of serum of allophenic mice, but not of any others, to the culture medium. The serum probably contains a factor (antigen–antibody complex?) that prevents the development of the immunologi-cal reaction *in vivo* (Wegmann *et al.*, 1971). These experiments raise the question of the true nature of tolerance of an organism to its own antigens: is it based on elimination of "forbidden" clones, or is it a special form of interaction between cells marked by an antigen and the lymphocytes competent with respect to it?

Conclusion

Considerable progress has been made in recent years in the simulation of immunological processes *in vitro*.

An essential condition for primary antibody synthesis to take place is the ability of the cells to interact intensively *in vitro*. This intensity is achieved in organ cultures by preservation of the natural packing of the cells, so that the conditions for interaction among them are adequate. The induction of a primary response in suspension cultures requires a high cell density and intensive mixing. These factors increase the likelihood that the necessary contacts will be made among cells of the various categories, and in particular between macrophages and lymphocytes.

The main difference between the dynamics of antibody synthesis during the primary and the secondary response *in vitro* and the dynamics *in vivo* is that antibody formation *in vitro* goes on longer. *In vitro*, there is evidently a disturbance of regulation that leads to termination of the immune response. The cellular mechanism of this disturbance may be either lengthening of the life span of individual plasma cells, a disturbance of the kinetics of development of their clones, or, finally, the onset of differentiation of additional precursors for antibody-synthesizing cells.

Models of the delayed hypersensitivity and transplantation immunity reac-tions that have been developed *in vitro* are important not only for experimental research, but also in the clinical field, for they can help with the diagnosis of autoimmune diseases, and they can be used to detect compatibility between donor and recipient when suitable partners are chosen for the transplantation of organs and tissues.

The primary immune response of the transplantation immunity type has been induced *in vitro* in living cells carrying transplantation antigens. The reaction, which is manifested as blast-transformation in mixed cultures, can be carried as far as resorption of target cells.

The various models that reproduce interaction between lymphocytes and target cells *in vitro* can be used to investigate the mechanism of transplantation immunity, which is difficult to undertake *in vivo*. For example, a problem that requires solution is the degree to which direct contact between the lymphocytes and the cells undergoing resorption is necessary. The data on this problem are difficult to reconcile at present. The simplest suggestion is that the mechanism of transplantation immunity differs depending on whether the antigenic differences are weak or strong. The situation is complicated, however, by the fact that in PHA-transformation, both processes are simulated: resorption of target cells takes place both after direct contact with the lymphocytes and also by the action of cytotoxic substances liberated into the medium. Obviously, further analysis of these processes will lead to a considerable increase in our knowledge of the mechanism of transplantation immunity.

It has been shown that complex interactions can take place in a population of sensitized immunocompetent cells in the presence of antigen, changes that result in marked changes in the structure of that population, and in particular in death of a certain category of cells, probably belonging to the stroma. These results, of course, lead to the further question whether similar changes take place in lymphoid tissue *in vivo* during immune reactions.

Some of the chief stages of the immune response, namely, the stage of identification of the antigens and the beginning of differentiation of the immunologically competent effector cells, are unquestionably the result of cooperative interaction in which different categories of lymphoid cells participate. These categories include macrophages, B cells, and different subpopulations of T cells. The simulation of cooperative interaction *in vitro* is an extremely useful technique in modern immunology.

There is a particular need at present to be able to prepare an antigen-reactive unit *in vitro* that will contain thymocytes and cells of bone marrow origin. It has not yet been possible to do this by explantation of a mixture of bone marrow and thymus cells from unsensitized donors. On the other hand, a primary immune response can easily be induced by mixing *in vitro* cells obtained from spleens of pairs of irradiated mice that have been injected with: (1) thymus cells + antigen and (2) bone marrow cells + antigen (Hartmann, 1971). Thus, it follows that the cooperative interaction among cells that is essential to formation of the antigen-reactive unit can readily be brought about *in vitro* if thymocytes and bone marrow cells that have been passed through the spleens of irradiated recipients are explanted. In this connection, it can be assumed that if thymocytes and bone marrow cells are used, the antigen-reactive unit must also contain stromal cells, which are added to the corresponding populations during their passage through the spleens of the irradiated recipients.

CHAPTER V

Cell Lines in Lymphoid and Hematopoietic Tissue

Hematopoietic tissue is a complex population system. It consists of dividing and mature cells that have differentiated in different ways. The number of cells forming the hematopoietic tissue is enormous, its composition is being constantly and rapidly renewed, and many cells (including dividing cells) can move about from place to place. An important principle governing the structure of hematopoietic tissue is that the constant loss of cells is made good by division of cells that are not those that perform functions of importance to the animal as a whole. Special precursor cells, not yet functionally mature, are the ones that divide. The variety of precursor cells is less extensive than the variety of mature cells. Throughout life, processes of differentiation are taking place in the hematopoietic tissue, during which the precursor cells choose the direction of their subsequent development. This choice, like the regulation of the number of the population, forms the basis of existence of the hematopoietic tissue as an orderly cell system and a component of the whole organism. As with any population of living organisms, the relationships among its members are of decisive importance for hematopoietic tissue. Although hematopoietic tissue, being within the organism, is exposed to powerful influences from the organism's components (some of these influences being warning signals destined for this tissue alone), interactions within the population itself are in fact the most important influences to which cells of hematopoietic tissue are exposed.

A clear example of these influences is the interaction between thymocytes and cells that originate from the bone marrow, a phenomenon that lies at the basis of the earliest and most significant stages of immunological reactions (J. F. A. Miller and Mitchell, 1967). A no less important cooperative interaction among cells takes place during differentiation of hematopoietic stem cells on transplantation into an irradiated animal. N. S. Wolf and Trentin (1968) demonstrated this interaction when they treated irradiated mice by injecting bone marrow. Foci consisting of clones of hematopoietic cells formed in the recipients' spleens and bones. Predominantly myeloid or erythroid hematopoiesis was observed, depending on the influence of the stroma. For instance, the ratio between erythroid and granulocytic colonies in the spleen was about 3:1, but in the bones, it was less than unity (0.5–0.7). If fragments of bone marrow stroma are first introduced into the spleens of irradiated mice, the foci formed on the transplanted stroma are mainly myeloid in character, whereas in the spleen, away from contact with the bone marrow stroma, they are erythroid in character. An important discovery in recent years was the method of cloning hematopoietic stem cells, with the formation of hematopoietic clone colonies in the spleens of irradiated mice (Till and McCulloch, 1961). Restoration of the hematopoietic tissue of the irradiated donor after transplantation of bone marrow cells into the animal takes place through the proliferation and differentiation of these cells, which constitute a self-maintaining cell line. The stem cell is polypotent with regard to its powers of differentiation in the direction of formation of erythroid, myeloid, or megakaryocytic cells.

There is a single ancestral cell for the hematopoietic and lymphoid series of differentiation. That there is is shown by the discovery of common chromosome markers, produced by irradiation (Barnes et al., 1959), in the lymphoid and hematopoietic cells of irradiated animals; it is also shown by regeneration of the lymphoid tissue of radiation chimeras when repopulated by hematopoietic cells contained in clones that are the progeny of hematopoietic stem cells (Trentin et al., 1967). Besides hematopoietic stem cells (polypotent cells capable of maintaining themselves for a time corresponding to the life span of the recipient), hematopoietic and lymphoid tissues also contain lines of specialized precursor cells. Their ability to maintain themselves is less than that of stem cells (e.g., about 6 months for mouse lymphoid cells), and the scope for their differentiation is restricted to a particular histogenetic series. Precursor cells of the myeloid series, with only some of the properties of the original hematopoietic stem cells, are capable of limited proliferation. It is these cells that are found on cloning in agar cultures. These precursor cells form large myeloid clone-colonies, containing several thousand cells, that persist for 10–12 days, and then degenerate.

A detailed study of these specialized "semistem" precursor cells is one of the most urgent problems at present under scrutiny, and it is being studied chiefly by in vitro methods.

The possibilities of differentiation of the hematopoietic stem cell are not limited to the formation of leukocytes and erythrocytes. This cell may also differentiate into macrophages.

Experiments using bone marrow chimeras with chromosome markers have shown that macrophages in a zone of aseptic inflammation belong to the donor, i.e., that they are of bone marrow origin (Goldman and Walker, 1962). The same is true of macrophages of the peritoneal exudate (Balner, 1963; Volkman, 1966), lung, spleen, and thymus (Virolainen, 1968).

That macrophages growing in cultures from individual hematopoietic colonies of mouse radiation chimeras have the chromosome marker of the donor indicates that they are the progeny of hematopoietic stem cells that are the ancestors of the hematopoietic colonies (see Chapter I).

Thus, in the adult organism, erythroid, myeloid, and lymphoid cells, as well as macrophages of the connective tissue, are based on a single line of hematopoietic stem cells and are ultimately renewed through it.

These extensive possibilities of differentiation of hematopoietic stem cells do not mean, however, that all the cells of hematopoietic tissue, still less all the cells of the tissues of the internal milieu, are the progeny of hematopoietic stem cells and are linked with them histogenetically in the adult organism. This is least clear in relation to the stromal cells of hematopoietic tissue and to other types of mechanocytes (fibroblasts, osteogenic cells, etc.). Determination of the range of histogenetically independent cell lines that together form the hematopoietic and lymphoid tissue and the entire system of tissues of the internal milieu is thus a problem that urgently requires solution. The functional unity of these tissue systems, on which the classic histologists such as Maksimov, Zavarsine, Ruymantsev, and Khlopin insisted, continues to receive fresh experimental confirmation. The question of the cell lines to which the stromal cells of hematopoietic tissue belong is particularly important in this connection.

During the first decades of work with tissue cultures, the transformation of cells from one form into another attracted particular attention. Many investigators studied the intermediate types of cells in the hope of obtaining information on cell transformations in culture. The problem that attracts the main attention of modern workers is whether the change in composition of the cell population *in vitro* is the result of cell transformation or selection. With the appearance of methods using chromosome markers, thymidine labeling, and cloning, new opportunities have been made available for the study of this problem. The search for morphological transformations in culture, with which investigators were concerned for many years, has now lost much of its urgency. The reason is, first, that morphologically identical cell forms are not homogeneous. For example, the population of small lymphocytes, with their identical morphology, is in fact a set of cells performing different functions: some are antigen-identifying lymphocytes that arise from the thymus, others are antigen-sensitive lymphocytes of bone marrow origin, a third group are perhaps hematopoietic stem cells, while a fourth group are effector lymphocytes responsible for the delayed hypersensitivity reaction. Second, and this is particularly important, the concentration of precursor cells that serve as the origin for particular lines of histogenesis in populations of hematopoietic and lymphoid cells is as a rule extremely low. For hematopoietic stem cells in the bone marrow, for instance, it is 10^{-3}; for antibody-synthesizing cells in the spleen, it is 10^{-4}; for precursor cells that form clones of fibroblasts in cultures of bone marrow, it is 10^{-5}. It is perfectly obvious that transformations of these cells are most unlikely to be detected by morphological observation. Such transformations are not characteristic and must be extremely rare events. A much more promising approach to the study of precursor cells during differentiation of hematopoietic and lymphoid cells is therefore by isolation of individual cell lines and analysis of their histogenetic possibilities.

It was shown above that the influence exerted by the stroma of hematopoietic and lymphoid organs on lymphopoiesis and hematopoiesis is clearly apparent both *in vivo* and in tissue culture. Whether the stromal cells and other categories of mechanocytes belong to the family of the hematopoietic stem cell or a special cell line (or lines) is thus of more than purely academic interest. Such matters as, for example, how to choose the optimal composition of the cell population used for the treatment of radiation sickness, and so on, depend largely on its solution. A cell line capable of prolonged self-maintenance has been isolated in monolayer cultures of hematopoietic tissue. Its cells, with the morphology of fibroblasts, have a diploid number of chromosomes and can be subcultured repeatedly. During this procedure, they retain the specific properties of the stroma of the hematopoietic organ the cells of which were used for explantation. For instance, fibroblasts grown in bone marrow cultures are capable of spontaneous osteogenesis if transplanted back into an animal, while fibroblasts from cultures of the spleen form reticular tissue. At the same time, cell lines of fibroblasts isolated from cultures of bone marrow, spleen, lymph nodes, thymus, and blood are indistinguishable in their morphology. They are clearly distinguishable from histiocyte—macrophages, however, even from those that are elongated and have a pair of polar processes. Judging by many of their features, fibroblasts develop in these cultures independently of histiocytes and show no histogenetic kinship with them.

It has been found that in established heterotopic transplants of hematopoietic and lymphoid tissue, stromal mechanocytes and their precursors are not replenished by hematopoietic and lymphoid cells of the recipient, even though all the hematopoietic and lymphoid cells are replaced by the host cells (Fridenshtein *et al.*, 1968). No transformation of hematopoietic cells into stromal mechanocytes has been observed in complete mice radiochimeras either. Stromal mechanocytes even of long-term radiochimeras are not replenished by donor cells (Fridenshtein and Kuralesova, 1971), in contrast to macrophages and hematopoietic and lymphoid cells. Thus, hematopoietic cells and stromal mechanocytes behave as two histogenetically independent cell lines both in radiochimeras and in heterotopic transplants.

When hematopoietic tissue from radiochimeras and from heterotopic transplants is explanted *in vitro*, fibroblasts that develop in culture also originate only from mechanocyte precursors and not from hematopoietic cells: in cultures of cells derived from heterotopic transplants, they have the donor's origin; in cultures of hematopoietic tissue cells from radiation chimeras, the recipient's origin. Macrophages in the same culture have the opposite origin (Fridenshtein, 1976).

The results cited seem to answer the question of the possibility of stromal mechanocyte recruitment both from hematopoietic cells and—negatively in this case—from macrophages.

As for other mechanocytes, the majority of the results obtained with cell markers show that they are also histogenetically independent of hematopoietic cells (for references, see Fridenshtein, 1976). It has been shown that connective tissue fibroblasts in canine radiochimeras are of recipient origin. These results are in agreement with data on cell identification in sarcomata induced in radio-chimeras by implantation of cellophane or by introduction of strontium-90. The same results were obtained in investigations in mice parabionts; in wound-healing, fibroblasts regenerate without participation of the partner cells, and sarcomata induced by implantation of polyvinyl chloride plates arise from local and not from circulating cells. PH chromosomes found in hematopoietic cells of patients with chronic myeloleukosis may serve as another marker. Macrophages in the patients' blood are also labeled with PH chromosome. Yet, attempts to find labeled fibroblasts proved unsuccessful; in all cases in which precise identi-fication of cell types was carried out, fibroblasts had no PH chromosomes.

Thus, contemporary methods fail to demonstrate the histogenetically com-mon nature of mechanocytes and hematopoietic cells either *in vitro* or *in vivo*. On the contrary, the results obtained rather indicate its absence. Precursors of mechanocytes are self-maintained independently of hematopoietic cells through-out postnatal life, and are not replenished at their expense. Despite an assump-tion made more than once in descriptive histology, neither lymphocytes nor macrophages, which are both descendants of hematopoietic stem cells, are transformed into mechanocytes.

Conclusion

Tissue cultures are nowadays being used on an ever-increasing scale to study problems connected with the structure, histogenesis, and functions of hemato-poietic and lymphoid tissue—problems that are not of purely theoretical interest, but are directly linked with current medical practice: treatment of disturbances of hematopoiesis, regulation of immunological functions, and transplantation of hematopoietic cells. With the use of tissue culture models, important informa-tion has been obtained on the cells, tissue structures, and interactions necessary for immunological processes to take place.

Tissue cultures provide the purest experimental system with which to study the action of substances that stimulate hematopoiesis (leukopoietins, erythro-poietins), as well as factors that inhibit hematopoietic tissue. Some important results have already been obtained in this respect, but the problem requires further analysis.

As a result of the development of the organ culture technique, it has been shown that a line of hematopoietic stem cells capable of supporting hemato-poiesis if transplanted back into irradiated recipients can be maintained *in vitro*. Organ cultures provide opportunities for *in vitro* study of factors that maintain and direct the differentiation of hematopoietic cells. The important role of tissue interactions in hematopoiesis can be judged from results already obtained. The character of these interactions during hematopoiesis is an important subject for future research. Finally, there is reason to hope that in the course of time, it will be possible to grow hematopoietic cells *in vitro* that will be suitable for transplantation under clinical conditions.

Recent Experience in Monolayer and Organ Cultures of Hematopoietic Tissue

E. A. Luriya and A. Ya. Fridenshtein

1. The Technique of Cloning Fibroblasts in Monolayer Cultures of Hematopoietic and Lymphoid Cells and Its Use to Study Stromal Precursor Cells

For further information on this subject, the reader is referred to the following references (some of which are cited in the text): Fridenshtein *et al.* (1970*b*, 1973*a,b*, 1974*a,b*, 1976), Luriya *et al.* (1972*b*), and Panasyuk *et al.* (1972).

Cloning of fibroblasts during primary explantation of hematopoietic and lymphoid tissue cells can be used as a method for the study of stromal precursor cells. Fibroblasts do not take part in hematopoiesis in the sense that they are not an intermediate stage in the differentiation of hematopoietic cells into mature cells, they do not form hematopoietic stem cells in the adult organism, and they do not arise from them. Nevertheless, stromal mechanocytes and their precursor cells are an important component of hematopoietic tissue responsible for the transfer of the specific microenvironment that acts on the territory of the hematopoietic and lymphopoietic organs (Fridenshtein *et al.*, 1974*a*).

Cloning of stromal fibroblasts can be carried out in monolayer cultures of bone marrow, embryonic liver, spleen, thymus, and lymph node cells. Cell suspensions are prepared by washing the cells out of these organs into Hanks' solution and then pipetting them or passing them from a syringe through needles of decreasing diameter. Before explantation, the cell suspensions are filtered through several layers of nylon. The cultures are grown in Roux flasks containing liquid culture medium. In cultures of this type, given the right initial density of explantation, discrete colonies of fibroblasts are formed, and by the 8th–10th day, they are big enough to be seen with the unaided eye. Colony formation begins on the 3rd or 4th day, when the colonies consist of a few cells. By the 10th day, the colonies are 0.5–0.8 cm in diameter and consist of several thousand cells. Between the 5th and 12th days of culture, the number of colonies does not increase; only the size of most of them does (Fridenshtein *et al.*, 1970, 1974*b*).

The cells that compose colonies are typical fibroblasts with low alkaline phosphatase activity and with numerous lysosomes rich in acid phosphatase. Characteristically, tonofibrils are found in their cytoplasm and a large nucleolar complex in their nuclei. The cells in the colonies synthesize collagen, which can be detected either histochemically or by the incorporation of labeled proline, in the form of hydroxyproline, into the proteins of the cells forming the colonies; these proteins are secreted into the medium.

No essential differences by which the fibroblasts forming colonies could be distinguished depending on the source of the explanted cells have yet been discovered. It has been found, though, that the structure of the colonies and the composition of the cells situated among the colonies and populating the culture in the period before colony formation distinguish cultures of different origin. During the first few days, mainly histiocytes are found in cultures of bone marrow and spleen cells: colonies of fibroblasts are formed in a more or less thick layer of these cells; after the 8th–10th day, the number of histiocytes diminishes. In cultures of thymus and lymph nodes, hardly any histiocytes are formed: after explantation, mainly lymphocytes can be seen on the surface of the glass, and after a few days, they die. From the 5th to the 7th days, against a background of degenerating cells, solitary large elongated cells, actively phagocytosing the cell debris, can be found; on the 8th–10th day, colonies of fibroblasts are formed, the space around them being almost free of cells. The structure of most colonies in 12–14 day cultures of bone marrow and thymus is usually compact, but looser in cultures of lymph nodes and spleen. Besides colonies consisting of fibroblasts, a few, evidently epithelial, colonies appear in cultures of the thymus.

To count the colonies, the cultures are fixed in 96° ethanol and stained by Giemsa's method. Most of the colonies are near average size for the given culture. However, depending on the shape of the colonies and the packing of the cells in them, considerable differences exist within each culture, a fact that evidently points to variation among the colony-forming cells within the same population. Most colonies have only one layer, but in large colonies, bands of fibroblasts often form many layers, and these bands are packed together particularly densely in the center of the colonies. Most fibroblasts in the colonies have the typical elongated shape and oval, pale nuclei. In addition, the colonies also contain very large cells with giant nuclei (evidently polyploid cells). About 2–5% of colonies consist of tile-shaped fibroblasts with denser nuclei.

The clonal nature of the colonies is supported by much evidence. The number of colonies formed has been shown to be a linear function of the number of explanted cells; typing the cells (by means of X and Y chromosomes) in colonies arising after explantation of a mixture of male and female cells has shown that all dividing cells within a colony are either male or female; finally, time-lapse cinematography of living cultures has yielded direct visual proof of the clonal nature of colonies.

Colony formation takes place only if explantation is carried out with a certain optimal initial density: if too many cells are explanted per unit surface area of the culture vessel, the fibroblasts grow out into a monolayer; if the density is too low, they do not grow at all.

The formation of discrete colonies of fibroblasts in cultures of hemato-poietic and lymphoid tissue cells is facilitated by the fact that on explantation, not only fibroblast colony-forming cells (FCFCs) are introduced into the culture, but also many other cells that play the role of natural feeders. These cells are lymphocytes and polynuclear cells that, on explantation, attach themselves to the surface of the glass—but die after a few days in culture—as well as monocytes, which are transformed into histiocyte–macrophages. Irradiated (4000 R) bone marrow cells also possess a complete feeder action for colony formation. Stable efficiency of colony formation (ECF), i.e., formation in which the number of colonies arising in cultures continues to be a linear function of the number of explanted cells, is achieved by explantation of intact lymphocytes or hematopoietic cells in an initial density between 10^4 and 10^5 cells/cm^2 or of irradiated bone marrow cells in a density between 5×10^5 and 10^6 cells/cm^2. The combined ECF for culture of a mixture of bone marrow cells from different donors with different ECF values is the mean of their ECFs when cultured separately; in the presence of standard feeder (irradiated bone marrow cells), the differences in ECF characteristic of these cell populations are reproduced. The addition of feeder cells in excess concentrations does not affect ECF and does not increase it above the level of stable effectiveness characteristic of explantation of bone marrow cells without the feeder cells. The addition of irradiated bone marrow cells also has a feeder action on the formation of colonies of fibroblasts in cultures of the thymus, spleen, and lymph glands when explanted in low initial density. Under these conditions, the stable ECF characteristic of these cells when explanted without the feeder, but in an adequate initial density, is also restored, but is not altered.

Taken as a whole, these findings show that the complete feeder action providing for stable ECF is ensured by the mere presence of a sufficient number of living (including irradiated) cells, and is not the privilege of a certain category of cells.

FCFCs are found in hematopoietic and lymphoid tissue of man, rabbits, guinea pigs, rats, and mice. For each of these species, the composition of the culture medium giving the highest ECF has been determined. For human, guinea pig, and rabbit cells, this is medium 199 or double-strength Eagle's medium; for mouse cells, it is Eagle's medium with the addition of the "nutrient cocktail" of Mishell and Dutton (1967). In every case, the medium must contain 10–20% fetal calf serum for guinea pig and mouse cells, and 10–20% homologous serum with the addition of 5% fetal calf serum for human and rabbit cells. The concentration of the precursors for fibroblasts is as follows: about 1–5 per 10^5 cells for the bone marrow of all the species of mammals mentioned above; 1 per 10^5 cells for the spleen of guinea pigs, rabbits, and mice; 1 per 10^6 cells for guinea pig, rabbit, and mouse thymus; and 1 per 10^7 cells for guinea pig lymphocytes. The cells concerned are precursor cells or mechanocytes (fibroblasts) with the ability to form clonal colonies of cells in monolayer cultures. Whether all the precursors for stromal mechanocytes present *in vivo* possess this last property, and consequently can be detected by the cloning method, is not yet known. Even for that category of precursors that form colonies in this way (FCFCs), however, the problem of the cloning efficiency, i.e., of what proportion of the FCFCs contained in the explant can give rise to colonies, has not been finally explained. According to some observations, cells that form colonies of fibroblasts under particular conditions of culture possess a high degree of resistance to harmful factors to which they are exposed on explantation. In fact, when the cells attached to the cover slip a few hours after explantation were removed with trypsin and transferred to another flask for further cultivation, practically all the colony-forming cells removed formed colonies after transfer.

It can be taken without qualification that the ECF reflects the concentration of a certain, but possibly not the only, category of precursors for fibroblasts existing among the cells of hematopoietic and lymphoid tissue. Cloning in monolayer cultures can be used to assess changes in the number of these cells under the influence of various factors that disturb the equilibrium state of hematopoietic tissue, to compare the numbers of these cells in different cell populations, and also to study the properties (e.g., radiosensitivity, behavior toward substances acting on the various phases of the cell cycle, and so on) of a given category of precursor cells. Two conditions must be observed: first, the work must be done within the stable range of ECF; second, parallel with explantation of the cell suspensions individually, they must be grown with the addition of standard feeder or as mixed cultures, if cells of the same type are to be compared, e.g., normal and treated bone marrow. This last condition ensures that the difference in ECF depends on a change in the FCFC concentration, not on a change in the feeder activity of the remaining cells present in the same populations.

The morphological identity of the colony-forming cells has not been precisely established. During the first few hours after explantation, they have the appearance of mononuclear cells; however, it is not certain whether they are reticular or monocytoid cells. After 24 hr, the cells forming colonies become elongated in shape, and they have an oval, pale nucleus with projecting nucleolus, which are features of fibroblast-like cells. They pass through their first S-period *in vitro* between 28 and 60 hr after explantation (Keilis-Borok *et al.*, 1971). These results were obtained by the autoradiographic study of cultures grown during the first 4–60 hr in medium containing thymidine-H^3 (1 μCi/ml medium) and then transferred for further growth to medium with unlabeled thymidine (25 mg/ml) to prevent reutilization of the label. Under these conditions, labeled fibroblasts were found in cultures to which thymidine-H^3 had been added for more than the first 28 hr; after cultivation with thymidine-H^3 for the first 60 hr, practically all the fibroblasts were labeled. It is interesting to note that in cultures to which thymidine-H^3 was added for less than 60 hr, the colonies of fibroblasts were either entirely labeled or entirely unlabeled, but never mixed, confirming the clonal nature of the colonies.

A high proliferating cell pool is maintained for a long time in the colonies: after saturation with thymidine-H^3 for 24 hr, the size of the pool was 100% in 7-day colonies, 89% in 10-day colonies, and 78% in 12-day colonies. To judge from the saturation curves, the mean fibroblast generation time in the colonies is about 20 hr; the time required for the number of fibroblasts to double in cultures between 8 and 14 days is about 26 hr (Keilis-Borok *et al.*, 1971; Panasyuk and Luriya, 1970).

Some properties of clongenic stromal cells were determined by the cloning method *in vitro*. Explantation of bone marrow cells from guinea pigs after saturation for 72 hr *in vivo* with thymidine-H^3, followed by investigation of histoautoradiographs of 3–5 day cultures (grown in a medium with the addition of unlabeled thymidine), was used to determine the percentage of FCFCs in a state of proliferation *in situ,* and consequently of cells that formed progeny in the cultures in the form of tritium-labeled fibroblasts (Epikhina and Latsinik, 1976; Epikhina, 1976). The proliferating pool in 6- and 14-day guinea pigs was found to contain 15% and 2%, respectively, of stromal colony-forming cells; in adult guinea pigs, practically all stromal colony-forming cells are outside the proliferating pool. That they are is also confirmed by the fact that these cells do not undergo thymidine suicide when incubated *in vitro* with thymidine-H^3 of high specific activity during the 30 min before explantation.

Stromal colony-forming cells have high adhesiveness. In the absence of serum, 90% of FCFCs from bone marrow and thymus adhere to the surface of the glass in 90 min; most, in fact, adhere to the surface of the glass during the first 30 min (Latsinik and Epikhina, 1973).

The content of FCFCs in the hematopoietic organs varies with age (Epikhina, 1976). For instance, the marrow of a single femur of a 10-day guinea pig contains about 4×10^3 FCFCs, or 30 FCFCs per 10^5 bone marrow cells; in guinea pigs aged 3 months, there are 19×10^3 FCFCs, or 20 FCFCs per 10^5 bone marrow cells; one femur of guinea pigs aged 1 year contains 2×10^3 FCFCs, or 2 FCFCs per 10^5 bone marrow cells.

The spleen of 10-day guinea pigs contains 0.7×10^3 FCFCs, or 7 FCFCs per 10^5 spleen cells; of 3-month guinea pigs, 2.5×10^3 FCFCs, or 1.5 FCFCs per 10^5 spleen cells; of guinea pigs aged 1 year, 2.3×10^3 FCFCs, or 0.5 FCFC per 10^5 spleen cells.

The thymus of 10-day guinea pigs contains 0.6×10^3 FCFCs, or 0.3 FCFC per 10^5 thymocytes; at 3 months, the guinea pig thymus contains 2×10^3 FCFCs, or 0.5 FCFC per 10^5 thymocytes; the number of FCFCs in the thymus of guinea pigs aged 1 year could not be determined.

The number of FCFCs in adult $(CBA \times C_{57}B1)F_1$ mice is 270, 220, and 175 in the marrow of one femur, in the thymus, and in the spleen, respectively (Fridenshtein *et al.*, 1976).

In contrast to macrophages, FCFCs belong to a cell lineage that is histogenetically independent of hematopoietic and lymphoid cells. That they do has been shown by typing colony-forming fibroblasts in cultures of bone marrow cells of radiation chimeras and bone marrow cells of semisyngenetic heterotopic bone marrow grafts. The typing was carried out by the indirect Coons' method using anti-H-2 antisera and fluorochrome-labeled rabbit serum against mouse immunoglobulins. These investigations showed that in complete bone marrow radiation chimeras in which hematopoietic bone marrow cells and also the macrophages formed in culture are donor's cells, the fibroblasts that form colonies are recipient's cells. Conversely, in 3-month semisyngenetic [from CBA to $(CBA \times C_{57}B1)F_1$] heterotopic bone marrow grafts transplanted under the capsule of the kidney, the bone marrow cells and also the macrophages formed in culture are recipient's cells, whereas fibroblasts in colonies formed in the cultures continue to be donor's cells (Ivanov-Smolenskii *et al.*, 1976).

The radiosensitivity of FCFCs has been determined by studying the inhibition of colony formation after irradiation of cell suspensions by different doses of X rays before explantation of the cells into monolayer cultures. The number of colonies in cultures of irradiated cells was compared in these experiments with the number of colonies in cultures of the same cells but not irradiated (Fridenshtein *et al.*, 1974b). The curve of survival of guinea pig and human bone-marrow FCFCs after irradiation *in vitro* is characterized by a D_0 of 178 R and n of 1.4. For mouse bone marrow FCFCs, $D_0 = 220$ R, but n is still 1.4.

Cells composing fibroblast colonies in monolayer cultures can easily be subjected to passage after separation from the glass by 0.25% trypsin solution; the passage can be repeated over and over again (as many as 20 passages in the case of fibroblasts from human and guinea pig bone marrow cultures). Such cultures preserve the diploid number of chromosomes, and they die out after 20 passages; i.e., they behave like diploid cell strains. If cells of an irradiated feeder are added for the passage, clone-colonies of fibroblasts are formed in the passages, just as in the primary culture. All this is evidence that colony-forming cells belong to the category of long self-maintaining precursor cells, possibly with stem cell properties.

Fibroblasts of hematopoietic tissue of adult mice tolerate passage with difficulty, and usually not more than two passages are possible. It is not yet clear whether this is due to their limited self-maintaining capacity or to other causes.

If 12–14 day cultures of bone marrow cells are incubated before passage for 1 hr with thymidine-H^3 of high specific activity, about one-third of the FCFCs undergo thymidine suicide; this finding indicates that the overwhelming majority of colony-forming cells present in colonies *in vitro* are in the proliferating pool (Epikhina, 1976).

Diploid strains of stromal fibroblasts grown in culture can be retransplanted *in vivo*. In the case of syngenetic or autotransplantation, cells grown *in vitro* can be transplanted as a thick cell suspension under the kidney capsule, whereas in allogenetic transplantation, they are introduced into diffusion chambers, made from millipore filters (pore size 0.22μ or $0.45\mu m$), which are then implanted intraperitoneally. After retransplantation of fibroblasts of bone marrow origin in diffusion chambers, bone tissue is formed, but if the material is transplanted under the kidney capsule, bone trabeculae with bone marrow cells surrounding them are found. If fibroblasts of splenic origin are transplanted in diffusion chambers, reticular tissue is formed, but if they are transplanted beneath the kidney capsule, reticular tissue infiltrated by lymphocytes is found (Fridenshtein *et al.,* 1974*a*). It is not yet known whether each clone of fibroblasts grown in cultures of spleen and bone marrow cells can transfer the appropriate microenvironment of the spleen or bone marrow. If rabbit fibroblasts of splenic, medullary, or thymic origin grown *in vitro* are added to cultures of spleen cells of normal rabbits set up by the method of Mishell and Dutton (1967) with the addition of SRBC, fibroblasts of thymic origin increase the number of antibody-forming cells detected by Jerne's method by 4–10 times (depending on the number of fibroblasts added). Addition of fibroblasts of bone marrow origin leads to a decrease in the number of antibody-producing cells; fibroblasts of splenic origin can have either an inhibitory or a stimulating action, depending on how many of them are added to the culture (Sidorenko *et al.,* 1975).

Several factors to which the organism is exposed can have an appreciable effect on the number of FCFCs in its hematopoietic and lymphoid organs (Fridenshtein *et al.,* 1974*b*). For instance, the number of FCFCs in the bone marrow of guinea pigs is increased threefold 2 hr after bleeding. After subcutaneous injection of an antigen (diphtheria toxoid) into guinea pigs, the number of FCFCs in the regional lymph node rises: by 30 times after 24 hr, by 40 times after 7 hr; it returns to its initial level after 1 month. In the contralateral regional lymph node, the number of FCFCs is also increased, but by a lesser degree and later. After curettage of the medullary cavity of one femur in mice and guinea pigs, the number of FCFCs increased not only in the curetted femur, but also in bone marrow in other areas (F. D. Wilson et al., 1974; Chailakhyan et al., in press).

Changes in the number of FCFCs in the bone marrow of mice and guinea pigs have also been studied after total and local irradiation in doses of 150–200 R. The results are reported in a special survey (Fridenshtein et al., 1976).

Colonies of fibroblasts can be grown not only in monolayer cultures, but also in agar bone marrow cultures and in cultures on a methylcellulose base. The cloning efficiency under these circumstances is lower, the colonies are smaller, and they are more difficult to maintain by passage. Such cultures have the advantage, however, that they can be used to analyze fibroblast and granulocyte precursors simultaneously (Metcalf, 1972).

2. The Method of Multiple Organ Cultures of Hematopoietic and Lymphoid Tissue

For further information on this subject, the reader is referred to the following references (some of which are cited in the text): Latsinik et al. (1970a) and Luriya et al. (1969b, 1971b,c, 1972b, 1974).

Among other methods that are used to culture lymphoid and hematopoietic cells, a special position is occupied by organ cultures. The advantage of organ cultures is that they are as yet the only method of in vitro cultivation by which it is possible to maintain myeloid hematopoiesis for a long time (for weeks or even months—Luriya et al., 1969b, 1971c; Latsinik et al., 1970; Chertkov et al., 1974) in cultures, including the proliferation of hematopoietic stem cells. When retransplanted from culture into an irradiated animal, these stem cells form hematopoietic colonies in the spleen. However, compared with results obtained by methods of cultivation of hematopoietic and lymphoid cells such as culture in semiliquid gel, in which cloning of myeloid (Pluznik and Sachs, 1965; Bradley and Metcalf, 1966), erythroid (McLeod et al., 1974), and megakaryocytic (Metcalf et al., 1975) precursors can be achieved, and compared with suspension cultures of lymphocytes in which antibody synthesis can be induced (Mishell and Dutton, 1966, 1967), the results obtained by organ culture are more difficult to assess quantitatively.

Various types of millipore filters, differing not only in pore size but also in certain surface properties, have been used as the supporting substrate for explants. The type of filter has a dramatic effect on the morphology of the tissue grown on it, on the formation of organotypical structures, and on the duration of hematopoiesis in the cultures. Of the filters that the authors have tested, the AUFS (0.6–0.9 μm) and RUFS (1.2 μm) filters (Chemapol, Czecho-slovakia) and the AAWP (0.8 μm) and RAWG (1.2 μm) filters (Millipore, United

States) have given good results for cultures of mouse embryonic liver. Hemato-poiesis has been kept going for the longest time (up to 2–2.5 months) by the use of AUFS and AAWP filters.

For explantation of human rib bone marrow, the type of filter has a decisive effect on the cell composition of the hematopoietic tissue. When AUFS filters are used, maturing myeloid cells in culture on a filter pass through it and are found in the nutrient medium, while chiefly immature myeloid cells remain on the filter. When VUFS filters (0.1–0.3 μm), which are impermeable to cells, are used, the composition of the hematopoietic cells is changed, comprising pre-dominantly differentiated forms (Kolesnikova, 1974).

Although cultivation of lymph nodes and thymus for a short period (2 weeks) is possible on HAWG filters (0.45 μm) that are impermeable to cells, filters with larger pores must be used for long-term cultivation: AUFS (0.6–0.9 μm), RUFS (1.2 μm), and RAWG (1.2 μm). It is important to note that although filters of the RUFS and RAWG types are similar with regard to the important parameter of pore size, the morphology of tissue growth on the filter and the degree to which cells penetrate into the interior of the filter and emerge on its other surface differ. When RAWG filters are used, practically no spread of cells on the lower surface of the filter is observed, and migration of cells into the interior of the filter is also negligible. The morphology of cultures on RUFS and also on AUFS (with smaller pores) filters is different: the cells actively grow into the filter and form compact bands on its under surface. It is not known what properties of the filters are responsible for this striking difference in the behavior of the cells.

Explants are applied to the filters as pieces of tissue measuring several millimeters.

During the formation of a multilayered zone of growth around the explanted fragment, in which mainly proliferation and differentiation of the hematopoietic and lymphoid cells located above the stromal tissue takes place, cells migrating from the explants onto the filter and coming into contact with its surface play an important role. The layer of cells formed on the surface of the filter determines the spread from the explants of other cell layers not yet in contact with the filter, but in contact with the surface of the cells, i.e., under conditions probably closer to those obtaining *in vivo*.

In this connection, it is interesting to examine how growth of the explant is affected by soaking the filters with a substance such as collagen, a physiological substrate for the cells. It has been found that soaking HAWG filters with collagen not only does not interfere with the organ culture of bone marrow fragments, but also actually provides better conditions for the formation of an extensive zone of growth from stromal fibroblasts: on HAWG filters not coated with collagen, the zone of growth is extremely small and consists of histiocytes. By combining filters with physiological substrates such as collagen, methods that provide a new approach to the optimization of the conditions of organ culture can thus be obtained.

The filters manufactured by the Chemapol firm and some types of filters manufactured by the Millipore firm are impregnated with a substance that may have a toxic action on cultures, especially if the material is explanted as a suspension of disconnected cells, rather than as a fragment of the tissue. Before use, the filters must therefore be boiled in two or three changes of double-distilled water.

Millipore filters are kept above the liquid nutrient medium by means of a transparent plastic platform. One of the vessels used for organ culture (a Conway dish), the plastic platforms, and diagrams showing certain alternative methods of setting up multiple organ cultures on small and large filters are illustrated in Figs. 3 and 4 (pp. 10 and 11).

A vessel that can be closed with a lid rendered airtight by means of a layer of a mixture of mineral oil and wax is best used in cases when the vessel is filled with a gas mixture (5–10% CO_2 in air) from a cylinder. If the composition of the gaseous phase can be kept constant by means of an incubator, plastic dishes or glass petri dishes can be used for organ culture. An important factor is setting the level of the nutrient medium: it must be a little below the surface of the platform on which the filter rests. It has a marked effect on the morphology of the cultures. Simultaneous cultivation of a large number of explants in the same vessel also gives better results. For instance, hematopoiesis is maintained longer when 10–12 fragments are explanted on a common nutrient medium than if only 4–6 fragments are used. With 10–12 fragments, better conditioning of the medium takes place; i.e., metabolic products of the explants are secreted into the medium and have a beneficial effect on the state of the tissue in culture. It is not known which component—stromal or hematopoietic—plays the leading role in conditioning the medium.

The choice of nutrient medium is determined by the material in culture. Besides chemically deficient media, biological media (serum or, in some cases, embryonic extract) have had to be introduced.

Better results with mouse embryonic liver cultures are obtained by the use of Parker's medium 199 in Earle's solution than by the use of Eagle's basic medium with the addition of vitamins and amino acid and NCTC-135 medium (Chertkov *et al.*, 1974). The medium 199 is supplemented with 20% bovine serum. Fetal calf serum is not recommended for use with such cultures, since it causes hematopoiesis to stop during the first weeks of culture. It is desirable to add vitamin C (7 mg/100 ml medium), glucose (400 mg/100 ml medium), L-gluta-mine (20 mg/100 ml medium), and extract of 12-day chick embryos to the medium. Addition of 10% embryonic extract is essential for the maintenance of hematopoiesis *in vitro*. Ready-prepared embryonic extract can be kept only in the frozen state, and it is thawed immediately before use.

The culture medium is changed every 48–72 hr, but the time can be varied depending on the intensity of growth of the cultures. The medium is renewed to the extent of half to two-thirds, but not completely, so as to preserve the conditioning factors.

The same nutrient medium, but with human group AB (IV) serum instead of bovine serum, has also been used for human bone marrow cultures. The addition of fetal calf serum to the medium did not inhibit hematopoiesis in cultures of human bone marrow.

In accordance with the early work of Fell, besides the components already mentioned, the authors have added sodium β-glycerophosphate to medium for osteogenic bone marrow tissue in a dose of 20 mg/100 ml medium.

Explants of hematopoietic and lymphoid tissue have been subjected to morphological and functional analysis. To investigate their morphology, sections or total preparations were made from the explanted tissue fragment and zone of growth, including the millipore filter, which becomes transparent after dehydration and treatment with xylol. Total preparations, mounted in Canada balsam, are most suitable for studying the morphology of the zone of growth, in which most important processes of differentiation of stromal tissue and maturation of hematopoietic and lymphoid cells take place in several layers.

It will be remembered that tissue mounted on a filter cannot be fixed with acetone, methanol, or other fixatives that dissolve millipore filters and thereby cause damage to the structure of the explant. The most commonly used fixatives are ethanol (80° and 96°), alcohol–formol, and 10% formalin.

When choosing a stain, it is important that the filter itself not take up the stain. To study culture morphology, various formulas including alum hematoxylin can be used. The authors have never succeeded in staining fixed cultures on filters with blood stains, for intense staining of the filter also occurred. The composition of lymphoid and hematopoietic cells from cultures is investigated in squash preparations or smears fixed with methanol and stained with azure–eosin by Romanovsky's method. To prepare smears, cells from cultures are suspended in a small volume of serum.

In addition to tissue growing on a filter, in some cases cells that have passed through the pores of the filter into the nutrient medium as well as cells attached to the bottom of the vessel have been analyzed. For instance, the medium from cultures of bone marrow fragments was centrifuged, and smears were made from the residue suspended in serum. In particular, by investigating cells of organ cultures of human bone marrow that had passed through the filter into the medium, it was possible to conclude that there is no promyelocytic block to differentiation *in vitro* (see p. 52), and that of cells kept in nutrient medium, about half are maturing myeloid cells (Table 13). For this reason, in order to obtain a true picture of the process of hematopoiesis in culture, data for the hematopoietic cells both on the filter and in the medium must be compared. It is

TABLE 13. Cells Found in Nutrient Medium of Organ Cultures of Human Bone Marrow[a]

Cell form	Cultivation time (days)			
	3	7	10	14
Neutrophils				
Promyelocytes	2.0	0	2.0	8.0
Myelocytes	10.0	9.0	11.0	19.0
Metamyelocytes	7.0	8.0	7.0	5.0
Stab cells	7.0	6.0	8.0	4.0
Segmented cells	59.0	53.0	44.0	33.0
Eosinophils				
Promyelocytes	0	0	0	0
Myelocytes	0	1.0	0	0
Stab cells	1.0	2.0	3.0	1.0
Segmented cells	4.0	9.0	14.0	18.0
Polychromatophilic normoblasts	4.0	2.0	5.0	0
Lymphocytes	2.0	3.0	0	1.0
Plasma cells	0	1.0	0	0
Reticular cells	4.0	6.0	6.0	8.0
Macrophages	0	0	0	3.0
Unidentified cells (per 100 neutrophils)	12.0	14.0	17.0	38.0

[a]After Kolesnikova et al. (1974).

possible to carry out a quantitative analysis of changes in the numbers themselves on the filter and in the medium.

When cultures are grown on filters permeable to cells, growth of colonies of fibroblasts can often be detected on the bottom of the vessel.

The ability of lymphoid and hematopoietic tissue to proliferate and differentiate *in vitro* has been studied by thymidine-H[3] labeling. Thymidine-H[3] (sp act 1 mCi/ml) was added to the nutrient medium in concentrations between 0.05 and 1.0 μCi/ml. If the material was examined 1–24 hr after addition of the isotope, thymidine-H[3] was used in a concentration of 1μCi/ml, but if the duration of culture with thymidine-H[3] was longer than 24 hr, the concentration was reduced to between 0.05 and 0.1 μCi/ml. Every day, further portions of thymidine-H[3] were added to the medium; this procedure is essential and adequate for the detection of all cells exhibiting proliferative activity over the period of the investigation.

Squash preparations and smears from cultures were fixed with methanol, treated with perchloric acid, washed with distilled water, dried, coated with photographic emulsion, exposed for 5–14 days, and developed by the usual method. In some cases, autoradiographs of serial sections of the cultures were obtained.

By labeling with thymidine-H^3, the composition of the proliferating pool of hematopoietic and lymphoid tissue can be determined *in vitro*, and the values of certain temporal parameters of differentiation of nondividing cells such as small lymphocytes, immature polymorphs, stab cells, and segmented polymorphs can be determined.

The cellular forms that comprise the proliferating pool in cultures can be deduced from the categories of hematopoietic tissue cells that are labeled when thymidine-H^3 is added to the medium 1 hr before fixation of 8-day cultures. In the case of mouse embryonic liver cultures, these cells were myeloblasts, promyelocytes, myelocytes, and some metamyelocytes (Luriya *et al.*, 1971*b*, 1974); in cultures of human bone marrow, they were myeloblasts and myelocytes (Luriya *et al.*, 1974). Differentiation of proliferating cells in cultures into mature, nondividing cells of the myeloid series, the rate of such differentiation, and the life span of the mature nondividing cells in cultures can be judged from the kinetics of appearance of labeled nondividing myeloid cells (stab cells and

TABLE 14. Kinetics of Proliferative Activity of Hematopoietic Stem Cells in Cultures of Mouse Embryonic Liver (According to Thymidine Suicide Values)[a]

Age of culture (days)	Incubation with thymidine-H^3	Number of cells injected per mouse	Number of colonies in spleen M±m	Thymidine-H^3 suicide (%)	P
0	−	80,000	10.0±0.45	37	<0.05
	+		6.0±0.4		
3	−	45,000	6.0±0.3	46	<0.05
	+		3.2±0.4		
7	−	40,000	8.6±0.5	53	<0.05
	+		4.1±0.3		
10	−	54,000	9.0±0.6	66	<0.05
	+		5.0±0.0		
14	−	60,000	16.0±1.2	45	<0.05
	+		9.7±0.4		
17	−	40,000	9.0±1.0	51	<0.05
	+		4.0±0.9		
24	−	15,000	4.8±0.6	37	<0.05
	+		3.0±0.2		
31	−	40,000	11.0±0.9	41	<0.05
	+		6.5±0.5		
40	−	40,000	15.8±0.6	33	<0.05
	+		10.0±0.8		
50	−	17,500	4.5±0.6	33	ND
	+		3.4±0.8		
60	−	24,000	1.6±0.3	11	ND
	+		1.4±0.3		

[a]After Rudneva (1974).

segmented polymorphs) in experiments in which 8-day cultures were kept in medium with thymidine-H^3 during the last 3, 4, 7, 9, 19, 24, 48, and 78 hr.

The stem potential of mouse hematopoietic tissue cultures is estimated by the method of Till and McCulloch (1961). Cells from cultures are washed off the filters with nutrient medium by repeated pipetting and injected into lethally irradiated syngenetic recipient mice at a dose of several tens or hundreds of thousands of cells ($n \times 10^3 - 10^4$) per recipient. Ten days later, the number of hematopoietic colonies in the fixed spleens is counted.

To study the kinetics of the CFUs in organ cultures of embryonic liver, it is best to grow the cultures on larger filters, on which many explants can be accommodated. The management of such cultures and the preparation of cells for transplantation are easier than in the case of cultures on small filters. Chertkov et al. (1974) showed that hematopoietic stem cells can be maintained for more than 2 months in cultures of the livers of 17-day CBA mouse embryos. Under these circumstances, as the use of the vinblastin technique and thymidine suicide showed, the CFUs in the cultures proliferated actively during the period of investigation (Table 14). Vinblastin was added to the medium 24 hr before the cells were removed, in a concentration of 1 μg/ml, as in the method of Bruchovsky et al. (1965). For thymidine suicide by the method of Becker et al. (1965), thymidine of high specific activity was used at a concentration of 100 μCi/ml.

When injected into lethally irradiated CBA recipients, CFUs from organ culture of CBA_{T6T6} mouse embryonic liver lead to formation of radiation chimeras: all dividing bone marrow, spleen, thymus, and lymph node cells carry the donor's chromosome marker (Chertkov et al., 1974).

Of the various methods of cultivation available, organ cultures evidently provide the simplest model of hematopoiesis in vitro, in which myelopoiesis resulting from proliferation of hematopoietic stem cells can be maintained for a long time.

References

Abelev, G. I., and Bakirov, R. D., 1967, The synthesis of embryonic serum antigens by the liver *in vitro, Vopr. Med. Khim.,* No. 4, 378.

Abelev, G. I., Perova, S. D., *et al.,* 1963, Production of embryonal α-globulin by transplantable mouse hepatomas, *Transplantation* 1:174.

Adler, W. H., Takaguchi, T., Marsh, B., and Smith, R. T., 1970, Cellular recognition by mouse lymphocytes *in vitro.* II. Specific stimulation by histocompatibility antigens in mixed cell culture, *J. Immunol.* 105:984.

Alfred, E. F., Corvazier, P., and Graber, P., 1963, Formation of antibody in tissue culture: Effect of certain mucopolysaccharides and hexosamine, *Nature London* 200:698.

Ambrose, C. T., 1964, Stimulation and inhibition by steroid hormones of the secondary response initiated *in vitro, Fed. Proc. Fed. Amer. Soc. Exp. Biol.* 23:556.

Ambrose, C. T., and Coons, A. H., 1963, Studies on antibody production. VIII. The inhibitory effect of chloramphenicol on the synthesis of antibody in tissue culture, *J. Exp. Med.* 117:1075.

Auerbach, R., 1965, Experimental analysis of lymphoid differentiation in the mammalian thymus and spleen, in: *Organogenesis* (De Haan and Ursprung, eds.), p. 539, Holt, Rinehart & Winston, New York.

Auerbach, R., and Globerson, A., 1965, Primary immune reactions in organ cultures, *Science* 149:991.

Auerbach, R., and Globerson, A. 1966, *In vitro* induction of the graft versus host reaction, *Exp. Cell Res.* 42:31.

Avrorov, P. P., and Timofeevskii, A. D., 1915, The cultivation of colorless blood corpuscles *in vitro, Russk. Vrach* 14:553.

Bain, B., Vas, M. R., and Lowenstein, L., 1964, The development of large immature mononuclear cells in mixed leukocyte cultures, *Blood* 23:108.

Ball, W. D., and Auerbach, R., 1960, *In vitro* formation of lymphocytes from embryonic thymus, *Exp. Cell Res.* 20:245.

Balner, H., 1963, Identification of peritoneal macrophages in mouse radiation chimeras, *Transplantation* 1:217.

Barnes, O. W. H., Corp, M. J., Ilbery, P. L. T., Loutit, J. F., *et al.,* 1959, Murine leukemia and the radiation chimeras, in: *Proceedings of the Third Canadian Cancer Conference*, p. 3, Academic Press, New York.

Becker, A. J., McCulloch, E. A., Siminovitch, L., and Till, J. E., 1965, The effect of differing demands for blood cell production on DNA synthesis by hematopoietic colony-forming cells of mice, *Blood* 26:296.

Bendinelli, M., and Wedderburn, N., 1967, Haemolytic plaque formation by unimmunized mouse peritoneal lymphocytes, *Nature London* 215:157.

Benevolenskaya, S. V., 1929, Haematopoiesis in cultures of the embryonic liver of man, *Arch. Exp. Zellforsch.* 9:128.

Bergel, R., 1930, Zur Wandlungsfähigkeit der Lymphozyten, *Arch. Exp. Zellforsch.* 9:269.

Berman, L., Stulberg, C. S., and Ruddle, F. H., 1955, Long-term tissue culture of human bone marrow, *Blood* 10:896.

Billen, D., 1957, Recovery of lethally irradiated mice by treatment with bone marrow cells maintained *in vitro, Nature London* 179:574.

Billen, D., 1959, Effect of bone marrow culture *in vitro* on its protective action in irradiated mice, *J. Natl. Cancer Inst.* 23:1389.

Biryuzova, V. I., and Kondratenko, V. G., 1964, Ultrastructural changes in irradiated cells of a bone marrow culture, *Radiobiologiya* 1:123.

Bloom, B. R., and Bennett, B., 1966, Mechanism of a reaction *in vitro* associated with delayed-type hypersensitivity, *Science* 153:80.

Bloom, W., 1927, Transformation of lymphocytes of thoracic duct into polyblasts (macrophages) in tissue cultures, *Proc. Soc. Exp. Biol. N.Y.* 24:567.

Bloom, W., 1928, Mammalian lymph in tissue culture. From lymphocyte to fibroblast, *Arch. Exp. Zellforsch.* 5:269.

Bloom, W., 1937, Transformation of lymphocytes into granulocytes *in vitro, Anat. Rec.* 69:99.

Bond, V. P., Fliedner, T. M., and Archambeau, J. O., 1965, *Radiation Death of Mammals. Disturbance of the Kinetics of Cell Populations*, Academic Press, New York and London.

Bradley, T. R., 1968a, Aspects of stimulation of bone marrow colony growth *in vitro, Aust. J. Exp. Biol. Med. Sci.* 46:335.

Bradley, T. R., 1968b, Stimulation of mouse bone marrow colony growth *in vitro* by conditioned medium, *Aust. J. Exp. Biol. Med. Sci.* 46:607.

Bradley, T. R., and Metcalf, D., 1966, The growth of mouse bone marrow cells *in vitro, Aust. J. Exp. Biol. Med. Sci.* 44:287.

Bradley, T. R., and Semienowicz, R., 1968, Colony growth of rat bone marrow cells *in vitro, Aust. J. Exp. Biol. Med. Sci.* 46:595.

Bradley, T. R., and Sumner, M. A., 1968, Stimulation of mouse bone marrow colony growth *in vitro* by conditioned medium, *Aust. J. Exp. Biol. Med. Sci.* 46:607.

Bradley, T. R., Metcalf, D., and Robinson, W., 1967, Stimulation by leukemic sera of colony formation in solid agar outlines by proliferation of mouse bone marrow cells, *Nature London* 213:926.

Brondz, B. D., 1968, Complex specificity of immune lymphocytes in allogeneic cell cultures, *Folia Biol. Prague* 14:115.

Bruchovsky, N., Owen, A., Becker, A. J., and Till, J. E., 1965, Effect of vinblastine on the proliferative capacity of L-cells and their progress through the division cycle, *Cancer Res.* 25:1232.

Brunner, K. T., Mauel, J., and Schlindler, R., 1966, *In vitro* studies of cell-bound immunity; cloning assay of cytotoxic action of sensitized lymphoid cells on allogeneic target cells, *Immunology* 11:499.

Bullock, W. W., Katz, D. H., and Benacerraf, B., 1975, Induction of T-lymphocyte responses to a small molecular weight antigen. III. T–T cell interactions to determinants linked together: Suppression vs. enhancement, *J. Exp. Med.* 142:275.

Bussard, A. E., and Hannoun, C., 1966, Antibody production by cells in tissue culture. II. Qualitative and quantitative aspects of antibody production (local hemolysis in gum) by cells obtained from long term culture, *J. Exp. Med.* **123**:1047.

Bussard, A. E., and Lurie, M., 1967, Primary antibody response *in vitro* in peritoneal cells, *J. Exp. Med.* **125**:873.

Bussard, A. E., Nossal, G. J., *et al.*, 1970, *In vitro* stimulation of antibody formation by peritoneal cells, *J. Exp. Med.* **131**:917.

Caron, G. A. .and Sarkany, I., 1966, Role of plasma factors in the transformation of peripheral blood lymphocytes into lymphoblasts, *Nature London* **210**:314.

Carpenter, R. R., 1963, *In vitro* studies of cellular hypersensitivity. I. Specific inhibition of migration of cells from adjuvant-immunized animals by purified protein derivative and other protein antigens, *J. Immunol.* **91**:803.

Carrel, A., 1923, A method for the physiological study of tissues *in vitro*, *J. Exp. Med.* **38**:407.

Carrel, A., and Burrows, M. T., 1910*a*, Culture de moelle osseuse et de rate, *C. R. Soc. Biol. Paris* **69**:299.

Carrel, A., and Burrows, M. T., 1910*b*, Cultivation of adult tissues and organs outside of the body, *J. Amer. Med. Assoc.* **55**:1379.

Carrel, A., and Ebeling, A. H., 1922, Pure cultures of large mononuclear leukocytès, *J. Exp. Med.* **36**:365.

Chailakhyan, R. K., 1970, Dynamics of formation of fibroblast-like cells in cultures of hematopoietic tissue, *Arkh. Anat.*, No. 3, 75.

Chailakhyan, R. K., Fridenshtein, A. Ya., and Vasil'ev, A. V., 1970, Clone formation in monolayer cultures of bone marrow and spleen, *Byull. Eksp. Biol. Med.*, No. 2, 94.

Chalmers, D. G., Coulson, A. S., Evans, C., and Yealland, S., 1966, Immunologically stimulated human peripheral blood lymphocytes *in vitro*. II. Mixed lymphocyte culture with related and unrelated donors, *Int. Arch. Allergy Appl. Immunol.* **30**:177.

Chan, S. H., and Metcalf, D., 1972, Local production of colony stimulating factor within the bone marrow. Role of non-hematopoietic cells, *Blood* **40**:646.

Chan, S. H., Metcalf, D., and Standley, E. R., 1971, Stimulation and inhibition by normal human serum of colony formation *in vitro* by bone marrow cells, *Br. J. Haematol.* **20**:329.

Chasovnikov, N., 1927, Cultivation of the thymus gland *in vitro*. Preliminary communication, *Arch. Exp. Zellforsch.* **3**:250.

Chen, J. M., 1954, The cultivation in fluid medium of organized liver, pancreas, and other tissue of fetal rats, *Exp. Cell Res.* **7**:518.

Chertkov, J. L. [Chertkov, I. L.], Samoylina, N. L. [Samoilina, N. L.], Rudneva, N. A., Rakcheev, A. M., and Kondratenko, N. F., 1974, Haemopoietic stem cell (CFU) kinetics in the culture, *Cell Tissue Kinet.* **7**:259.

Clemmesen, H., Espersen, T., and Plum, C. M., 1948, *In vitro* study of bone marrow. III. Erythropoiesis *in vitro* of sternal marrow from cases of pernicious anemia and lymphatic leucosis under therapy, *Blood* **3**:155.

Cosenza, H., and Leserman, L. D., 1972, Cell interactions in antibody formation *in vitro*. I. Role of third cell in the *in vitro* response of spleen cells to erythrocyte antigens, *J. Immunol.* **108**:418.

Dantschakoff, W., 1924, Wachstum transplantierter embryonaler Gewebe in der Allantois, *Z. Anat. Entwicklungsgesch.* **74**:102.

David, J. R., 1966, Delayed hypersensitivity *in vitro*: Its mediation by cell-free substances formed by lymphoid cell–antigen interaction, *Proc. Natl. Acad. Sci. U.S.A.* **56**:72.

David, J. R., Al-Ashari, S., and Lawrence, H. S., 1964, Delayed hypersensitivity *in vitro*. I. The specificity of the inhibition of cell migration by antigens, *J. Immunol.* **93**:264.

De Bruyn, P. P. H., 1945, The motion of migrating cells in tissue cultures of lymph nodes, *Anat. Rec.* 93:295.

Dexter, T. M., and Lajtha, L. G., 1974, Proliferation of haemopoietic stem cells *in vitro*, *Br. J. Haematol.* 28:525.

Dick, K. A., and Van Bekkum, D. W., 1972, Evidence for the identity of CFU*a* and CFUs, in: *In Vitro Culture of Hematopoietic Cells*, p. 136, Publication of the Radiobiological Institute TNO, Rijswijk.

Didukh, M. S., and Fridenshtein, A. Ya., 1970, Histogenetic relations between lymphocytes and reticulum cells in lymph gland transplantation, *Tsitologiya* 7:901.

Didukh, M. S., and Luriya, E. A., 1967, Regeneration of irradiated lymph nodes after transplantation *in vitro* and *in vivo*, *Arkh. Anat.*, No. 6, 33.

Diener, E., and Armstrong, W. D., 1967, Induction of antibody formation and tolerance *in vitro* to a purified protein antigen, *Lancet* 2:1281.

Diener, E., and Feldman, M., 1972, Mechanism at the cellular level during induction of high zone tolerance *in vitro*, *Cell. Immunol.* 5:130.

Dutton, R. W., and Mishell, R. I., 1967, Cell populations and cell proliferation in the *in vitro* response of normal mouse spleen to heterologous erythrocytes, *J. Exp. Med.* 126:443.

Eagle, H., 1955, Nutrition needs of mammalian cells in tissue culture, *Science* 122:501.

Elves, M. W., Israels, M. C., and Collinge, M., 1966, An assessment of the mixed leucocyte reaction in renal failure, *Lancet* 1:682.

Epikhina, S. Yu., 1976, Characteristics of stromal precursor cells of hematopoietic and lymphoid tissue detected by their cloning *in vitro*, Candidate's dissertation, Moscow.

Epikhina, S. Yu., and Latsinik, N. V., 1976, Proliferative activity of stromal bone marrow precursor cells with clonogenic properties, *Byull. Eksp. Biol. Med.*, No. 1, 55.

Erdmann, R., 1917, Cytological observations on the behavior of chicken bone marrow in plasma medium, *Amer. J. Anat.* 22:73.

Evans, R., and Alexander, P., 1970, Cooperation of immune lymphoid cells with macrophages in tumor immunity, *Nature London* 228:620.

Farnes, R., and Trobaugh, F. E., 1961*a*, The inhibitory effect of collagenase on bone marrow fibroblasts *in vitro*, *Exp. Cell Res.* 24:612.

Farnes, P., and Trobaugh, F. E., 1961*b*, Observations on leukemic marrow explants in well cultures, *J. Lab. Clin. Med.* 57:568.

Fazekas, J., Bacsy, E., and Rappay, G., 1968, *In vitro* cultivation of rat bone marrow, *Haematol. Hung.* 2:349.

Fazzari, I., 1926, Culture *in vitro* di milzi embrionale e d'adulta, *Arch. Exp. Zellforsch.* 2:307.

Feldman, M., and Bleiberg, I., 1967, Studies on the feedback regulation of hemopoiesis, in: *Ciba Found. Symp. Cellular Differentiation*, London, p. 79.

Fell, H. B., and Robinson, R., 1929, The growth, development and phosphatase activity of embryonic avian femora and limb buds cultivated *in vitro*, *Biochem. J.* 23:767.

Fischer, A., 1930, Gewebezuchtung, in: *Handbuch der Biologie der Gewebezellen in Vitro*, Müler & Steinike, Munich.

Fisher, M., and Doljanski, L., 1929, Über das Wachstum von Milzstromazellen *in vitro*, *Arch. Entwicklungsmech. Org.* 116:123.

Fishman, M., 1959, Antibody formation in tissue culture, *Nature London* 183:1200.

Fishman, M., 1961, Antibody formation *in vitro*, *J. Exp. Med.* 114:837.

Foot, N. C., 1912, Über das Wachstum von Knochenmark *in vitro*. Experimenteller Beitrag zur Entstehung des Fettgewebe, *Beitr. Pathol. Anat.* 53:446.

Foster, R., Metcalf, D., and Kirchmyer, R., 1968, Induction of bone marrow colony-stimulating activity by a filterable agent in leukemic and normal mouse serum, *J. Exp. Med.* 127:853.

Fridenshtein, A. Ya., 1963, Experimental extracellular osteogenesis [in Russian], Meditsina, Moscow.

Fridenshtein, A. Ya. [Friedenstein, A. J.], 1973, Determined and inducible osteogenic precursor cells, in: *Hard Tissue Growth, Repair and Remineralization, Ciba Found. Symp.*, p. 169.

Fridenshtein, A. Ya., 1976, Precursor cells of mechanocytes, *Int. Rev. Cytol.*, p. 327.

Fridenshtein, A. Ya. [Friedenstein, A. J.], and Kuralesova, A. I., 1971, Osteogenic precursor cells of bone marrow in radiation chimeras, *Transplantation* 12:99.

Fridenshtein, A. Ya. [Friedenstein, A. J.,], and Lalykina, K. S., 1972, Thymus cells are inducible to osteogenesis, *Eur. J. Immunol.* 2:602.

Fridenshtein, A. Ya. [Friedenstein, A. J.], and Luriya, E. A. [Luria, E. A.], 1974, Differenzierung von lymphatischem und hämatopoetischem Gewebe in Organkulturen, *Fortschr. Haematol.* 3:342.

Fridenshtein, A. Ya., Petrakova, K. V., Kuralesova, A. I., *et al.*, 1968, Heterotopic transplants of bone marrow, *Transplantation* 6:230.

Fridenshtein, A. Ya., Chailakhyan, R. K., and Lalykina, K. S., 1970a, Fibroblast-like cells in cultures of guinea pig hematopoietic tissue, *Tsitologiya* 9:1147.

Fridenshtein, A. Ya. [Friedenstein, A. J.], Chailakhyan, R. K. [Chailakhjan, R. K.], and Lalykina, K. S., 1970b, The development of fibroblast colonies in monolayer cultures of guinea pig bone marrow and spleen cells, *Cell Tissue Kinet.* 3:393.

Fridenshtein, A. Ya., Luriya, E. A. Chailakhyan, R. K., *et al.*, 1970c, Hematopoietic tissue in organ and monolayer cultures *in vitro*, *Vestn. Akad. Med. Nauk SSSR*, No. 7, 86.

Fridenshtein, A. Ya., Chailakhyan, R. K., Latsinik, N. V., Panasyuk, A. F., and Keilis-Borok, I. V., 1973a, Stromal cells responsible for the microenvironment in hematopoietic and lymphoid tissues: Cloning *in vitro* and retransplantation *in vivo*, *Probl. Gematol. Pereliv. Krovi*, No. 10, 14.

Fridenshtein, A. Ya., Deriglazova, Yu. V., and Kulagina, N. N., 1973b, The formation of fibroblast colonies in monolayer cultures of bone marrow, spleen, thymus, and peritoneal and pleural cells, *Byull. Eksp. Biol. Med.*, No. 10, 90.

Fridenshtein, A. Ya. [Friedenstein, A. J.], Chailakhyan, R. K., Latsinik, N. V., Panasyuk, A. F. [Panasuk, A. F.], and Keilis-Borok, I. V. [Keiliss-Borok, I. V.], 1974a, Stromal cells responsible for transferring the microenvironment of the hemopoietic tissue, *Transplantation* 17:331.

Fridenshtein, A. Ya. [Friedenstein, A. J.], Deriglazova, Yu. F. [Deriglasova, U. F.], Kulagina, N. N., Panasyuk, A. F. [Panasuk, A. F], Rudakova, S. F. [Rudakowa, S. F], Luriya, E. A. [Luria, E. A.], and Rudakov, I. A. [Rudakow, I. A.], 1974b, Precursors for fibroblasts in different populations of hematopoietic cells as detected by the *in vitro* colony assay method, *Exp. Hematol. Copenhagen* 2:83.

Fridenshtein, A. Ya., Gorskaya, U. F., and Kulagina, N. N., 1976, Fibroblast precursors in normal and irradiated mouse hematopoietic organs, *Exp. Hematol. Copenhagen* 4:276.

Friedman, H. P., Stavitsky, A. B., and Solomon, J. M., 1965, Induction *in vitro* of antibodies to phage T_2 : Antigens in the RNA extract employed, *Science* 149:1106.

Gallien-Lartique, O., 1966, Action du facteur erythropoietique du plasma sur l'hémapoièse *in vitro* du foie de souris embryonnaire, *Exp. Cell Res.* 41:109.

Galustyan, Sh., 1940, Morphology of the thymus in the light of experimental analysis, Doctoral dissertation, Leningrad.

Gey, G. O., 1933, An improved technique for massive tissue culture, *Amer. J. Cancer* 17:752.

Ginsburg, H., 1968, Graft versus host reaction in tissue culture. I. Lysis of monolayers of embryonal mouse cells from strains differing in H-2 histocompatibility locus by rat lymphocytes sensitized *in vitro*, *Immunology* 14:621.

Gitlin, D., Kitzes, J., and Boesman, M., 1964, Cellular distribution of serum α-foetoprotein in organ of the foetal rat, *Nature London* 215:543.

Globerson, A., and Auerbach, R., 1966, Primary antibody response in organ cultures, *J. Exp. Med.* 124:1001.

Globerson, A., and Auerbach, R., 1967, Reaction *in vitro* of immunocompetence in irradiated mouse spleen, *J. Exp. Med.* 223:1967.

Goldman, A. S., and Walker, B. E., 1962, The origin of cells in the infiltrates found at the sites of foreign protein injection, *Lab. Invest.* 11:808.

Govaerts, A., 1960, Cellular antibodies in kidney homotransplantation, *J. Immunol.* 85:516.

Gowans, J. L., 1957, The effect of the continuous re-infusion of lymph and lymphocytes on the output of lymphocytes from the thoracic duct of unanesthetized rats, *Br. J. Exp. Pathol.* 38:67.

Granger, G. A., and Weiser, R. S., 1964, Homograft target cells: Specific destruction *in vitro* by contact interaction with immune macrophages, *Science* 145:1427.

Granger, G. A., Shacks, D. J., Williams, T. W., *et al.*, 1969, Lymphocyte *in vitro* cytotoxicity: Specific release of lymphotoxin-like materials from tuberculin-sensitive lymphoid cell, *Nature London* 221:1155.

Grobstein, S., 1953, Morphogenetic interaction between embryonic mouse tissues separated by membrane filter, *Nature London* 172:869.

Grobstein, S., 1956, Trans-filter induction of tubules in mouse metanephrogenic mesenchyme, *Exp. Cell Res.* 10:424.

Gurvich, A. E., and Sidorova, E. V., 1964, A study of inhibition of antibody synthesis, *Biokhimiya*, No. 3, 556.

Gurvich, A. E., Drizlikh, G. I., Sidorova, E. V., and Tumanova, A. E., 1965, Role of repressive factors in antibody biosynthesis, in: *Molecular and Cellular Bases of Antibody Formation*, p. 515, Czechoslovakian Academy of Sciences, Prague.

Halpern, B., Storb, U., and Fray, A., 1967, Delayed hypersensitivity *in vitro*, *Nature London* 215:400.

Hannoun, C., and Bussard, A. E., 1966, Antibody production by cells in tissue culture. I. Morphological evolution of lymph node and spleen cells in culture, *J. Exp. Med.* 123:1035.

Harris, G., and Littleton, R. J., 1966, The effects of antigens and of phytohemagglutinin on rabbit spleen cells suspensions, *J. Exp. Med.* 124:621.

Harrison, R. G., 1907, Observations on the living developing nerve fiber, *Proc. Soc. Exp. Biol. New York* 4:140.

Hartmann, K. U., 1971, Induction of a hemolysin response *in vitro*, *J. Exp. Med.* 133:1323.

Haskill, J. S., and Marbrook, J., 1972a, In vitro immunity to sheep erythrocytes by fractionated spleen cells: Differentiation within the antibody-forming cell precursor population, *Cell. Immunol.* 3:448.

Haskill, J. S., and Marbrook, J., 1972b, In vitro response of partially purified spleen cells: "Blast" transformation, *Cell. Immunol.* 5:93.

Haskill, J. S., McNeill, T. A., and Moore, M. A., 1970, Density distribution analysis of *in vivo* and *in vitro* colony forming cells in bone marrow, *J. Cell. Physiol.* 75:167.

Häyry, P., Andersson, L. C., Nordling, S., and Virolainen, M., 1972, Allograft response *in vitro*, *Transplant. Rev.* 12:91.

Hellström, I., and Hellström, K. E., 1973, Cellular immunity and blocking serum activity in chimeric mice, *Cell. Immunol.* 7:73.

Hellström, I., Hellström, K. E., Evans, C. A., Heppner, G. H., *et al.*, 1969, Serum-mediated protection of neoplastic cells from inhibition by lymphocytes immune to their tumor-specific antigens, *Proc. Natl. Acad. Sci. U.S.A.* 62:362.

Hellström, K. E., and Hellström, I., 1967, Allogeneic inhibition and its relationship to cell-bound immunity phenomena, in: *First International Congress of the Transplantation Society* (abstracts), p. 84, Paris.

Hersh, E. M., and Harris, J. E., 1968, Macrophage–lymphocyte interaction in antigen-induced blastogenic response of human peripheral blood leukocytes, *J. Immunol.* 100:1184.

Hillis, W. D., and Bang, F. B., 1962, The cultivation of human embryonic liver cells, *Exp. Cell Res.* 26:9.

Holm, G., 1966, *In vitro* cytotoxic effects of lymphoid cells from rats with experimental autoimmune nephrosis, *Clin. Exp. Immunol.* 1:45.

Holm, G., and Perlmann, P., 1965, Phytohaemagglutinin-induced cytotoxic action of unsensitized immunologically competent cells on allogeneic and xenogeneic tissue culture cells, *Nature London* 202:818.

Holm, G., and Perlmann, P., 1967, Quantitative studies on phytohaemagglutinin-induced cytotoxicity by human lymphocytes against homologous cells in tissue culture, *Immunology* 12:525.

Holm, G., and Perlmann, P., 1969, Cytotoxicity of lymphocytes and its suppression, in: *The Immune Response and Its Suppression*, Vol. 15, p. 295, S. Karger, Basel.

Holtermann, O. A., and Nordin, A. A., 1968, Primary induction of plaque-forming antibody-producing cells in spleen organ culture, *Proc. Soc. Exp. Biol. N. Y.* 127:675.

Huggins, C., 1931, The formation of bone under the influence of epithelium of the urinary tract, *Arch. Surg.* 22:377.

Ichikawa, Y., Pluznik, D. H., and Sachs, L., 1967, Feedback inhibition of the development of macrophages and granulocyte colonies. I. Inhibition by macrophages, *Proc. Natl. Acad. Sci. U.S.A.* 58:1480.

Iscove, N. N., and Sieber, F., 1975, Erythroid progenitors in mouse bone marrow detected by macroscopic colony formation in culture, *Exp. Hematol.*Copenhagen 3:32.

Ivanov-Smolenskii, A. A., Gorskaya, U. F., Kuralesova, A. I., and Latsinik, N. V., 1976, The origin of stromal mechanocytes in the bone marrow cell culture, *Byull. Eksp. Biol. Med.* No. 10, p. 1270.

Jacobson, L. O., Marks, E. K., and Gaston, E. O., 1959, Studies on erythropoiesis. XII. The effect of transfusion-induced polycythemia in the mother of the fetus, *Blood* 14:647.

Jacoby, F., 1965, Macrophages, in: *Cells and Tissues in Culture*, Vol. 2, p. 1, Academic Press, London and New York.

Katz, D. H., and Benacerraf, B., 1972, The regulatory influence of activated T cell on B cell responses to antigens, *Adv. Immunol.* 15:2.

Keilis-Borok, I. V., Latsinik, N. V., and Epikhina, S. Yu., 1971, Dynamics of formation of fibroblast colonies in monolayer cultures of bone marrow from data of thymidine-H^3 incorporation *in vitro*, *Tsitologiya* 11:1402.

Keilis-Borok, I. V., Latsinik, N. V., and Deriglazova, Yu. F., 1972, Characteristics of bone marrow precursor cells for fibroblast-like cells with respect to their incorporation of thymidine-H^3, *Byull. Eksp. Biol. Med.*, No. 10, 91.

Kenneth, N. N., 1971, Histological changes in long-term explants of human lymph nodes during lymphoblastoid transformation, *Acta Pathol. Microbiol. Scand.* 79:243.

Klein, R., 1958, Étude des fonctions lymphocytaires par la cinématographie en contraste de phase de cultures de tissue humains foetaux, *C. R. Acad. Sci. Paris* 152:265.

Klein, R., 1959, Démonstration par la microcinématographie en contraste de phase de la transformation reversible des lymphocytes en macrophages *in vitro*, *C. R. Soc. Biol. Paris* 153:545.

Kokorin, I. N., Safronova, L. D., Miskarova, E. D., *et al.*, 1970, Functional and morphological characteristics of diploid strains of reticulum cells, *Vestn. Akad. Med. Nauk. SSSR*, No. 7, 57.

Kolesnikova, A. I., 1974, Hematopoiesis in normal and leukemic human bone marrow cultures, Candidate's dissertation, Moscow.

Kolesnikova, A. I., Luriya, E. A., Vorob'ev, A. I., Gordeeva, A. A., and Domracheva, E. V., 1974, Characteristics of hematopoiesis in organ cultures of human bone marrow, *Probl. Gematol. Pereliv. Krovi*, No. 11, 7.

Kozinets, G. I., 1962, A study of the proliferative ability of hematopoietic cells with the aid of radioactive indicators, *Probl. Gematol. Pereliv. Krovi*, No. 11, 41.

Krantz, S. B., 1968, Application of the *in vitro* erythropoietic system to the study of the human bone marrow disease polycythemia vera, in: Erythropoietin, *Ann. N. Y. Acad. Sci.* 149(art. U):430.

Krantz, S. B., and Goldwasser, E., 1965, On the mechanism of erythropoietin-induced differentiation. IV. Some characteristics of erythropoietin action on hemoglobin synthesis in marrow cell culture, *Biochim. Biophys. Acta.* 108:455.

Krontovskii, A., 1925, Pathologisch–physiologische Beobachtungen über Herzexplantate, *Arch. Exp. Zellforsch.*, No. 1, 58.

Kuz'menko, G. N., and Fridenshtein, A. Ya., 1971, Formation of fibroblast colonies in monolayer cultures of the spleen of irradiated and normal mice, *Byull. Eksp. Biol. Med.*, No. 1, 74.

Kuz'menko, G. N., Panasyuk, A. F., Fridenshtein, A. Ya., and Kulagina, N. N., 1972, Radiosensitivity of bone marrow cells forming colonies of fibroblasts in monolayer cultures, *Byull. Eksp. Biol. Med.*, No. 10, 94.

Lajtha, L. G., 1965, Bone marrow in culture, in: *Cells and Tissues in Culture*, Vol. 2 (E. N. Willmer, ed.), p. 173, Academic Press, London.

Lalykina, K. S., and Fridenshtein, A. Ya., 1969, Induction of osteogenesis in populations of lymphoid cells in guinea pigs, *Byull. Eksp. Biol. Med.*, No. 6, 105.

Latsinik, N. V., and Epikhina, S. Yu., 1973, Adhesive properties of cells of hematopoietic and lymphoid tissue forming colonies of fibroblasts in monolayer cultures, *Byull. Eksp. Biol. Med.*, No. 12, 86.

Latsinik, N. V., and Keilis-Borok, I. V., 1971, A study of fibroblast-like cells in monolayer cultures of mouse embryonic liver, *Byull. Eksp. Biol. Med.*, No. 6, 96.

Latsinik, N. V., Luriya, E. A., Samoilina, N. A., Fridenshtein, A. Ya., and Chertkov, I. L., 1969, Colony-forming cells in organ cultures of embryonic liver, *Byull. Eksp. Biol. Med.*, No. 7, 88.

Latsinik, N. V., Luria, E. A., [Luriya, E. A.], Fridenshtein, A. Ya. [Friedenstein, A. J.], Samoilina, N. L. [Samoylina, N. L.], and Chertkov, I. L. [Chertkov, J. L.], 1970a, Colony forming cells in organ cultures of embryonal liver, *J. Cell Physiol.* 75:163.

Latsinik, N. V., Samoilina, N. L., and Chertkov, I. L., 1970b, Erythropoietin-sensitivity of hematopoietic cells of embryonic liver during cultivation *in vitro*, *Probl. Gematol. Pereliv. Krovi*, No. 6, 31.

Latta, J. S., and Johnson, H. M., 1934, Studies of lymphatic tissue grown *in vitro* with splenic extract as culture medium, *Arch. Exp. Zellforsch.* 16:221.

Lau, P., Brody, J. I., and Beizer, L. H., 1967, *In vitro* development of bone marrows from patients with neutropenia, *Blood* 29:462.

Leserman, L. D., Cosenza, H., and Roseman, J. M., 1972, Cell interactions in antibody formation *in vitro*, *J. Immunol.* 109:587.

Lewis, M. R., 1925, The formation of macrophages, epitheloid cells, and giant cells from leukocytes in incubated blood, *Amer. J. Pathol.* 1:91.

Lewis, W. H., and Webster, L. T., 1921, Migration of lymphocytes in plasma cultures of human lymph nodes, *J. Exp. Med.* 33:261.

Lonai, P., Wekerle, H., and Feldman, M., 1972, Fractionation of specific antigen-reactive cells in an *in vitro* system of cell-mediated immunity, *Nature London New Biol.* 235:60.

Lucarelli, C., Procellini, A., Garnevali, C., *et al.,* 1968, Fetal and neonatal erythropoiesis, *Ann. N. Y. Acad. Sci.* 149:544.

Luriya, E. A., 1966, Organ cultures of the thymus and lymph glands in medium with heterologous serum, *Dokl. Akad. Nauk SSSR* 171(6):1431.

Luriya, E. A., and P'yanchenko, I. E., 1966, Cultivation of guinea pig lymph glands in organ cultures on medium with homologous serum, in: *Proceedings of Symposia on General Immunology of the N. F. Gamaleya Institute of Epidemiology and Microbiology* [in Russian], p. 55, Publication of the N. F. Gamaleya Institute, Moscow.

Luriya, E. A., and P'yanchenko, I. E., 1967, Cultivation of guinea pig lymph glands in organ cultures on medium with homologous serum, *Byull. Eksp. Biol. Med.,* No. 6, 81.

Luriya, E. A., and Snegireva, A. E., 1966, Effect of the epithelial stroma of the thymus on differentiation of lymphocytes, *Byull. Eksp. Biol. Med.,* No. 5, 103.

Luriya, E. A., Chakhava, O. V., and Fridenshtein, A. Ya., 1966*a* Origin of fibroblast-like cells in a bone marrow culture, *Tsitologiya* 1:115.

Luriya, E. A., Nikolaeva, A. I., Gurvich, A. E., and Fridenshtein, A. Ya., 1966*b,* Induction of antibody synthesis in organ cultures of lymph glands, in: *Proceedings of Symposia on General Immunology of the N. F. Gamaleya Institute of Epidemiology and Microbiology* [in Russian], p. 120, Publication of the N. F. Gamaleya Institute, Moscow.

Luriya, E. A., Nikolaeva, A. I., Gurvich, A. E., and Fridenshtein, A. Ya., 1967, Induction of antibody synthesis in organ cultures of lymph glands, *Byull. Eksp. Biol. Med.,* No. 7, 73.

Luriya, E. A., Bakirov, R. D., Abelev, G. I., and Fridenshtein, A. Ya., 1969*a,* Organ cultures of embryonic liver synthesizing serum proteins, *Byull. Eksp. Biol. Med.,* No. 3, 95.

Luriya, E. A., Bakirov, R. D., Eliseeva, T. A., *et al.,* 1969*b,* Differentiation of hepatic and hematopoietic cells and synthesis of blood serum proteins in organ cultures of the liver, *Exp. Cell Res.* 54:111.

Luriya, E. A., P'yanchenko, I. E., and Fridenshtein, A. Ya., 1969*c,* Organ cultures of hematopoietic tissue (embryonic liver), *Probl. Gematol. Pereliv. Krovi,* No. 5, 32.

Luriya, E. A. [Luria, E. A.], Panasyuk, A. E., and Fridenshtein, A. Ya., [Friedenstein, A. V.], 1971*a,* Fibroblast colony formation from cultures of blood cells, *Transfusion* 11:345.

Luriya, E. A., Panasyuk, A. F., and Prusevich, T. O., 1971*b,* Differentiation of hematopoietic cells in organ cultures, *Byull. Eksp. Biol. Med.* No. 3, 89.

Luriya, E. A., [Luria, E. A.], Samoilina, N. L., [Samoylina, N. L.], Gerasimov, Yu. V., [Gerasimow, U. V.], and Chertkov, I. L. [Chertkov, J. L.], 1971*c,* Proliferation of hematopoietic stem cells in cultures of mouse embryonic liver, *Byull. Eksp. Biol. Med.,* No. 4, 103 [in Russian]; *J. Cell Physiol.* 78:461 [in English].

Luriya, E. A., Panasyuk, A. F., Kuz'menko, G. N., and Fridenshtein, A. Ya. [Friedenstein, A. J.], 1972*a,* Effect of tuberculin and Freund's adjuvant on the formation of fibroblast colonies by immunocompetent cell populations, *Cell. Immunol.* 3:133.

Luriya, E. A., Vorob'ev, A. I., Kulagina, N. N., *et al.,* 1972*b,* Organ cultures of human bone marrow, *Probl. Gematol. Pereliv. Krovi,* No. 2, 6.

Luriya, E. A., Prusevich, T. O., and Kolesnikova, A. I., 1974, Proliferation and differentiation of hematopoietic and lymphoid tissue in organ cultures, *Probl. Gematol. Pereliv. Krovi,* No. 11, 13.

Maksimov, A. A., 1909, Untersuchungen über Blut und Bindegewebe. I. Die Frühesten Entwicklungsstadien der Blut und Bindegewebezellen beim Säugetierembryo, bis zum Angang der Blutbildung in der Leber, *Arch. Mikrobiol.* 73: 44.

Maksimov, A. A., 1916, Cultivation of connective tissue in adult mammals *in vitro, Arkh. Anat.* 1:105.

Maksimov, A. A., 1917, Sur la production artificielle des myélocytes dans les cultures de tissu lymphoide, *C. R. Soc. Biol. Paris* **80**:235.

Maksimov, A. A., 1922, Untersuchungen über Blut und Bindegewebe. VII. Ü ber *in vitro* Kulturen von lymphoidem Gewebe des erwachsenenn Säugetierorganismus, *Arch. Mikrosk. Anat.* **96**:494.

Maksimov, A. A., 1923*a*, Untersuchungen über Blut und Bindegewebe. VIII. Die zytologischen Eigenschaften der Fibroblasten, Retikulumzellen und Lymphozyten des lymphoiden Gewebes ausserhalb des Organismus ihre genetischen Wechselbeziehungen und prospektiven Entwicklungspotenzen, *Arch. Mikrosk. Anat.* **97**:283.

Maksimov, A. A., 1923*b*, Untersuchungen über Blut und Bindegewebe. IX. Über die experimentelle Erzeugung von myeloiden Zellen in Kulturen des lymphoiden Gewebes, *Arch. Mikrosk. Anat.* **97**:314.

Maksimov, A. A., 1925, Tissue cultures of young mammalian embryos, *Contrib. Embryol. Carnegie Inst. Washington* **16**:47.

Maksimov, A. A., 1927, Bindegewebe und blutbildende Gewebe, in: *Handbuch der mikroskopischen Anatomie des Menschen*, Vol. 2 (Möllendorff, ed.), p. 232, Springer-Verlag, Berlin.

Maksimov, A. A., 1928, Cultures of blood leukocytes. From lymphocyte and monocyte to connective tissue, *Arch. Exp. Zellforsch.* **5**:169.

Maksimov, A. A., 1929, Über die Entwicklung argyrophiler und kollagener Fasern in Kulturen von erwachsenem Säugetiergewebe, *Z. Mikrosk.-Anat. Forsch.* **17**:625.

Marbrook, J., 1967, Primary immune response in cultures of spleen cells, *Lancet* **2**:1279.

McArthur, W. P., Hutchinson, A. J., and Freeman, M. Y., 1969, Primary antiprotein antibody synthesis elicited *in vitro*, *Nature London* **221**:83.

McCulloch, E. A., and Parker, R. C., 1957, Continuous cultivation of cells of hemic origin, *Canadian Cancer Conference*, Vol. 2, p. 152, Academic Press, New York.

McCulloch, E. A., and Till, J. E., 1970, Cellular interaction in the control of hemopoiesis, in: *Hemopoietic Cellular Proliferation* (F. Stohlman, ed.), p. 15, Grune & Stratton, New York.

McCulloch, E. A., and Till, J. E., 1971, Effects of short-term culture on populations of hemopoietic progenitor cells from mouse marrow, *Cell Tissue Kinet.* **4**:11.

McKenna, J. M., 1961, Antibody protein synthesis *in vitro*, *Ann. N. Y. Acad. Sci.* **94**:131.

McLaurin, B. P., 1965, Homograft interaction in the test-tube, *Lancet* **2**:816.

McLeod, D. L., Shreeve, M. M., and Axelrod, A. A., 1974, Improved plasma culture system for production of erythrocytic colonies *in vitro:* Quantitative assay method for CFU-E, *Blood* **44**:517.

McWhorter, J., and Whipple, A. O., 1912, The development of the blastoderm of the chick *in vitro*, *Anat. Rec.* **6**:12.

Merrill, J. P., Hanan, C., and Hawes, M. D., 1960, A demonstration of a cytotoxic effect *in vitro* following the rejection of skin grafts by the rabbit, *Ann. N. Y. Acad. Sci.* **87**:266.

Metcalf, D., 1972, Formation in agar of fibroblast-like colonies by cells from the mouse pleural cavity and other forces, *J. Cell Physiol.* **80**:409.

Metcalf, D., and Foster, R., 1967, Behavior on transfer of serum stimulated bone marrow colonies, *Proc. Soc. Exp. Biol. N. Y.* **126**:758.

Metcalf, D., and Moore, M. A. S., 1971, *Haemopoietic Cells*, North Holland Publishing Co., Amsterdam.

Metcalf, D. and Moore, M. A. S., 1973, Regulation of growth and differentiation in haemopoietic colonies growing in agar, in: *Haemopoietic Stem Cells, Ciba Found. Symp.* **13**(*New Ser.*):157.

Metcalf, D., and Osmond, G., 1966, A radioautographic investigation of the identity of phytohaemagglutinin responsible cells in the lymphoid tissue of the rat, *Exp. Cell Res.* 41:669.

Metcalf, D., Bradley, T. R., and Robinson, W., 1967, Analysis of colonies developing *in vitro* from mouse bone marrow cells stimulated by kidney feeder layers of leukemic serum, *J. Cell Physiol.* 69:93.

Metcalf, D., McDonald, H. R., Odartchenko, N., and Sordat, B., 1975, Growth of mouse megakaryocyte colonies *in vitro*, *Proc. Natl. Acad. Sci. U.S.A.* 72:1744.

Michaelides, M. C., 1957, Antibody production in tissue culture, *Fed. Proc. Fed. Amer. Soc. Exp. Biol.* 16:426.

Michaelides, M. C., and Coons, A. H., 1963, Studies on antibody production. V. The secondary response *in vitro*, *J. Exp. Med.* 117:1035.

Micklem, H. S., and Loutit, J. F., 1966, *Tissue Grafting and Radiation*, Academic Press, New York and London.

Miller, J. F. A., and Mitchell, G. F., 1967, The thymus and the precursors of antigen reactive cells, *Nature London* 216:659.

Miller, J. F. A., Basten, A., Sprent, J., and Cheers, C., 1971, Interaction between lymphocytes in immune responses, *Cell. Immunol.* 2:469.

Miller, R. G., and Phillips, R. A., 1969, Separation of cells by velocity sedimentation, *J. Cell. Physiol.* 73:191.

Mishell, R. I., and Dutton, R. W., 1966, Immunization of normal mouse spleen cell suspensions *in vitro*, *Science* 153:1004.

Mishell, R. I., and Dutton, R. W., 1967, Immunization of dissociated spleen cell cultures from normal mice, *J. Exp. Med.* 126:423.

Miskarova, E. D., Lalykina, K. S., Kokorin, I. N., and Fridenshtein, A. Ya., 1970, Osteogenic potential of long-term diploid cultures of bone marrow cells, *Byull. Eksp. Biol. Med.*, No. 9, 78.

Miura, Y., Mizoguchi, H., Tacucu, F., and Nacao, K., 1968, *In vitro* effect of erythropoietin on the spleen of polycythemic mouse. III. Limited exposure to erythropoietin and actinomycin D, *Blood* 30:433.

Möller, E., 1965, Antagonistic effects of humoral isoantibodies of the *in vitro* cytotoxicity of immune lymphoid cells, *J. Exp. Med.* 122:11.

Monit, B., and Sato, G. H., 1967, Improved *in vitro* survival of normal, functional spleen cells, *Science* 157:449.

Moore, G. E., 1975, Cell lines from humans with hematopoietic malignancies, in: *Human Tumor Cells in Vitro* (J. Fogh, ed.), p. 299, Plenum Press, New York and London.

Moore, M. A., McNeil, T. A., and Haskill, J. S., 1970, Density distribution analysis of *in vivo* and *in vitro* colony forming cells in developing fetal liver, *J. Cell. Physiol.* 75:181.

Moscona, A., 1952, Cell suspensions from organ rudiments of chick embryos, *Exp. Cell Res.* 3:535.

Mosier, D. E., 1967, A requirement for two cell types in antibody formation *in vitro*, *Science* 158:1573.

Mosier, D. E., and Coppleson, L. W., 1968, A three-cell interaction required for the induction of the primary immune response *in vitro*, *Proc. Natl. Acad. Sci. U.S.A.* 61:542.

Mountain, I. M., 1955, Antibody production by spleen *in vitro*. I. Influence of cortisone and other chemicals, *J. Immunol.* 74:270.

Murray, R. G., 1947, Pure cultures of rabbit thymus epithelium, *Amer. J. Anat.* 81:369.

Nossal, G. J., 1958, Antibody production by single cells, *Br. J. Exp. Pathol.* 39:544.

Nossal, G. J., 1959, Antibody production by single cells. II The difference between primary and secondary response, *Br. J. Exp. Pathol.* **40**:118.

Nossal, G. J., 1960, Antibody production by single cells. IV. Further studies on multiply immunized animals, *Br. J. Exp. Pathol.* **41**:89.

Nossal, G. J., 1966, Antibody production in tissue culture, in: *Cells and Tissues in Culture,* Vol. 3 (E. N. Willmer, ed.), p. 317, Academic Press, London.

Nossal, G. J., and Lederberg, J., 1958, Antibody production by single cells, *Nature London* **181**:1419.

O'Brien, T. F., and Coons, A. H., 1963, Studies on antibody production. VII. The effect of 5-bromodeoxyuridine on the *in vitro* anamnestic antibody response, *J. Exp. Med.* **117**:1063.

O'Brien, T. F., Michaelides, M. C., and Coons, A. H., 1963, Studies on antibody production. VI. The course, sensitivity, and histology of the secondary response *in vitro, J. Exp. Med.* **117**:1053.

Oettgen, H. F., Silber, R., Mieschner, P. A., and Hirschhorn, R., 1966, Stimulation of human tonsillar lymphocytes *in vitro, Clin. Exp. Immunol.* **1**:77.

Oppenheimer, J. J., Wang, J., and Frei, E., 1965, The effect of skin homograft rejection on recipient and donor mixed leukocyte cultures, *J. Exp. Med.* **122**:651.

Ortiz-Muniz, G., and Sigel, M.M., 1967, Long-term synthesis of antibody *in vitro, Proc. Soc. Exp. Biol. N. Y.* **124**:1178.

Osgood, E. E., and Muscovitz, A. P., 1936, Culture of human bone marrow. Preliminary report, *J. Amer. Med. Assoc.* **106**:1888.

Panasyuk, A. F., Luriya, E. A., Fridenshtein, A. Ya., *et al.*, 1972, Cultures of fibroblast-like cells from human bone marrow, *Probl. Gematol. Pereliv. Krovi,* No. 1, 34.

Pappenheimer, A. A. M., 1913, Further studies of the histology of the thymus, *Amer. J. Anat.* **14**:299.

Paran, M., Sachs, L., Barak, Y., and Resnitzky, P., 1970, *In vitro* induction of granulocyte differentiation in hematopoietic cells from leukemic and non-leukemic patients, *Proc. Natl. Acad. Sci. U.S.A.* **67**:1542.

Perlmann, P., Perlmann, H., and Holm, G., 1968, Cytotoxic action of stimulated lymphocytes on allogeneic and autologous erythrocytes, *Science* **160**:306.

Pierce, C. W., 1969, Immune responses *in vitro.* I. Cellular requirements for the immune response by nonprimed and primed spleen cells *in vitro, J. Exp. Med.* **130**:345.

Pinkel, D., 1964, Cultivation of mouse thymus in organ culture, *Proc. Soc. Exp. Biol. N. Y.* **116**:54.

Playfair, J. H., Papermaster, B. W., and Cole, L. J., 1965, Focal antibody production by transferred spleen cells in irradiated mice, *Science* **149**:998.

Pluznik, D. H., and Sachs, L., 1965, The cloning of normal "mast" cells in tissue culture, *J. Cell. Comp. Physiol.* **66**:319.

Pluznik, D. H., and Sachs, L., 1966, The induction of normal "mast" cells by a substance from conditioned medium, *Exp. Cell Res.* **43**:553.

Popov, N. V., 1927, The histogenesis of the thymus as shown by tissue cultures, *Arch. Exp. Zellforsch.* **4**:395.

Poureau-Schneider, N., 1962, Sur la culture organotypique *in vitro* de la moelle osseuse de poule et de rat adultes, *C. R. Soc. Biol. Paris* **156**:1225.

Poureau-Schneider, N., Bernard, C., Boiron, M., *et al.*, 1963, Comportement de la moelle osseuse humaine normale et leucémique associée au rein embryonnaire de poulet en culture organotypique, *Exp. Cell Res.* **32**:51.

Prusevich, T. O., and Luriya, E. A., 1969, Maintenance of hematopoiesis in bone marrow grafted on a previously grown bone stroma, *Byull. Eksp. Biol. Med.,* No. 8, 96.

Prusevich, T. O., and Luriya, E. A., 1972, Proliferation and differentiation of lymphocytes in organ cultures of the neonatal mouse thymus, *Byull. Eksp. Biol. Med.*, No. 9, 98.

Pulvertaft, R. J., 1958, The effects of reduced oxygen tension on platelet formation *in vitro*, *J. Clin. Pathol.*, No. 11, 535.

Pulvertaft, R. J., 1959, Cellular associations in normal and abnormal lymphocytes, *Proc. R. Soc. Med.* **52**:315.

Pulvertaft, R. J., and Humble, J. G., 1956, Culture de moelle osseuse sur lames tournantes, *Rev. Hematol.* **11**:349.

Pulvertaft, R. J., and Jayne, H. W., 1953, Agar culture on exudates, *Ann. R. Coll. Surg. Engl.* **12**:161.

Rabinowitz, Y., 1963, Fresh serum adherence factor and the glass chromatographic separation of the leukocytes. Viability studies of purified cells including their development in tissue culture, *Blood* **22**:840.

Rachmilewitz, M., and Rosin, A., 1944, Studies on bone marrow *in vitro*. II. The effect of hemoglobin and red cell stromata in explanted bone marrow, *Amer. J. Med. Sci.* **208**:193.

Raff, M. C., and Wortis, H. H., 1970, Thymus dependence of theta-bearing cells in the peripheral lymphoid tissues in mice, *Immunology* **18**:931.

Rangan, S. R. S., 1967, Origin of the fibroblastic growth in chicken buffy coat macrophage cultures, *Exp. Cell Res.* **46**:477.

Rapp, F., and Melnick, J. L., 1966, Cell, tissue and organ cultures in virus research, in: *Cells and Tissues in Culture*, Vol. 3, p. 263, Academic Press, New York and London.

Rappay, G., Fazekas, I., and Gyevai, A., 1968, Effect of hydrocortisone on growth *in vitro* of fibroblast-like (FL) cells derived from rat organs, *Acta Med. Acad. Sci. Hung.* **25**:451.

Rasmussen, H., 1933, Über das Verhalten von Knochenmark in Gewebskulturen, *Arch. Exp. Zellforsch.* **14**:285.

Reisner, E. H., 1959, Tissue culture of bone marrow, *Ann. N. Y. Acad. Sci.* **77**:487.

Reisner, E. H., 1966, Tissue culture of bone marrow. II. Effect of steroid hormones on hematopoiesis *in vitro*, *Blood* **27**:460.

Reisner, E. H., 1967, Tissue culture of bone marrow. III. Myelostimulatory factors in serum of patients with myeloproliferative diseases, *Cancer* **20**:1679.

Rhoads, C. P., and Parker, F., 1928, Observations on incubated normal bloods, *Amer. J. Pathol.* **4**:271.

Richter, K. M., 1958, Studies on the maintenance of functional and anatomic organ-integrity in culture, *J. Okla. State Med. Assoc.* May, p. 252.

Reike, W. O., 1966, Lymphocytes from thymectomized rats: Immunologic, proliferative and metabolic properties, *Science* **152**:535.

Robinson, W. A., Marbrook, J., and Diener, E., 1967, Primary stimulation and measurement of antibody production to sheep red blood cells *in vitro*, *J. Exp. Med.* **126**:347.

Rode, H. N., and Gordon, J., 1970, The mixed leukocyte culture: A three component system, *J. Immunol.* **104**:1453.

Rogers, J. C., 1976, Identification of an intracellular precursor to DNA excreted by human lymphocytes, *Proc. Natl. Acad. Sci. U.S.A.* **73**:3211.

Rosenau, W., and Moon, H. D., 1961, Lysis of homologous cells by sensitized lymphocytes in tissue culture, *J. Natl. Cancer Inst. Washington* **27**:471.

Rosenau, W., and Moon, H. D., 1962, The inhibitory effect of hydrocortisone on lysis of homologous cells by lymphocytes *in vitro*, *J. Immunol.* **89**:422.

Rosenau, W., and Moon, H. D., 1966, Studies on the mechanism of the cytolytic effect of sensitized lymphocytes, *J. Immunol.* **96**:80.

Rosenthal, A. S., and Shevach, E. M., 1973, Function of macrophages in antigen recognition by guinea pig T lymphocytes. I. Requirement for histocompatible macrophages and lymphocytes, *J. Exp. Med.* **138**:1194.

Rubin, A. L., Stenzel, K. H., Hirschorn, K., *et al.*, 1964, Histocompatibility and immunologic competence in renal homotransplantation, *Science* **143**:815.

Ruddle, F. H., Berman, L., and Stulberg, C. S., 1958, Chromosome analysis of five long-term cell culture populations derived from non-leukemic human peripheral blood (Detroit strains), *Cancer Res.* **18**:1048.

Ruddle, N. H., and Waksman, B. H., 1968, Cytotoxicity mediated by soluble antigen and lymphocytes in delayed hypersensitivity, *J. Exp. Med.* **128**:1237.

Rudneva, N. A., 1974, Proliferative activity of hematopoietic stem cells in long-term cultures of mouse embryonic liver, *Byull. Eksp. Biol. Med.*, No. 8, 83.

Rumyantsev, A. V., 1932, *Tissue Culture in Vitro and Its Importance in Biology* [in Russian], Gos. Med. Izdat., Moscow.

Sabin, F. R., 1921, Studies on blood; vitally stainable granules as a specific criterion for erythroblasts and differentiation of three strains of white blood cells as seen in living chick's yolk-sac, *Bull. Johns Hopkins Hosp.* **32**:314.

Sainte-Marie, J., 1966, Cytokinetics of antibody formation, *J. Cell. Physiol.* **67**:109.

Salvatorelli, G., 1966, Observations sur l'hematopoiese *in vitro* dans la moelle osseuse embryonnaire de poulet, *C. R. Acad. Sci. Paris* **262**:666.

Salvatorelli, G., 1967*a*, Action des extraits de levure et de foie sur l'erythropoiese medullaire *in vitro* chez l'embryon de poulet, *C. R. Acad. Sci. Paris* **264**:344.

Salvatorelli, G., 1967*b*, Réactivation de l'erythropoiese de la moelle osseuse de poulet *in vitro* par l'apport de foie embryonnaire, *C. R. Acad. Sci. Paris* **265**:1219.

Saunders, G. C., and King, D. W., 1966, Antibody synthesis initiated *in vitro* by paired explants of spleen and thymus, *Science* **151**:1390.

Schlesinger, M., 1972, Antigens of the thymus, *Prog. Allergy* **16**:214.

Schmidtke, J. R., and Unanue, E. R., 1971, Macrophage–antigen interaction: Uptake, metabolism and immunogenicity of foreign albumin, *J. Immunol.* **107**:331.

Schwarz, M. R., 1966, Blastogenesis in cultures containing genetically dissimilar thymic cells, *Immunology* **10**:281.

Shekhter, S. Yu., 1965, Cultivation of bone marrow in suspension form for studying erythropoietic activity of plasma, *Pathol. Fiziol.*, No. 2, 81.

Shelton, E., and Rice, M. E., 1959, Growth of normal peritoneal cells in diffusion chambers: A study in cell modulation, *Amer. J. Anat.* **105**:281.

Sheridan, J., and Stanley, E. R., 1971, Tissue sources of bone marrow colony stimulating factor, *J. Cell. Physiol.* **78**:451.

Shiomi, C., 1925, Explantationsversuche mit Lymphknoten auf Plasma unter Zusatz von Milz-, Nevennieren und Knochenmarksextrakt unter Nachprüfung der Versuche von Maximow und unter besonderer Berücksichtigung der Bildung granulierter Zellen, *Virchows Arch.* **257**:714.

Shreffler, D. C., and David, C. S., 1975, The H-2 major histocompatibility complex and the I immune response region: Genetic variation, function and organization, *Adv. Immunol.* **21**:125.

Sidorenko, A. V., Korukova, A. A., Grigor'eva, O. G., and Gurvich, A. E., 1975, Effect of fibroblasts from monolayer cultures of hematopoietic and lymphoid tissues on the immune response *in vitro*, *Byull. Eksp. Biol. Med.*, No. 12, 1465.

Simonsen, M., 1957, The impact of the developing embryo and newborn animal of adult homologous cells, *Acta Pathol. Microbiol. Scand.* **40**:480.

Skvortsov, V. T., and Gurvich, A. E., 1968, Relative rates of synthesis of immunoglobulins and light chains in rabbit spleen cells during secondary response, *Nature London* **218**:377.

Sorieul, E., 1966, Aspect de l'hématopoiese en culture du foie du souris, *Eur. J. Cancer* 2:245.

Spadafina, L., 1935, Contributo allo studio delle culture *in vitro* di midollo osseo, *Arch. Exp. Zellforsch.* 17:43.

Spratt, N. T., 1947, A simple method for explanting and cultivating early chick embryos *in vitro*, *Science* 106:452.

Stanley, E. R., 1972, Physico-chemical characteristics of colony stimulating factors, in: *In Vitro Culture of Hemopoietic Cells*, p. 26, Publication of the Radiobiological Institute TNO, Rijswijk.

Stanley, E. R., and Metcalf, D., 1971, The molecular weight of colony stimulating factor, (CSF), *Proc. Soc. Exp. Biol. N. Y.* 137:1029.

Stanley, E. R., Robinson, W. A., and Ada, G. L., 1968, Properties of the colony stimulating factor in leukemic and normal mouse serum, *Aust. J. Exp. Biol. Med. Sci.* 46:715.

Stavitsky, A. B., 1961, *In vitro* studies of the antibody response, in: *Advances in Immunology*, Vol. 1, p. 211, Academic Press, New York and London.

Steiner, D. F., and Anker, H. S., 1956, On the synthesis of antibody protein *in vitro*, *Proc. Natl. Acad. Sci. U.S.A.* 42:580.

Stephenson, J. R., Axelrod, A. A., McLean, D. L., *et al.*, 1971, Induction of colonies of hemoglobin-synthesizing cells by erythropoietin *in vitro*, *Proc. Natl. Acad. Sci. U.S.A.* 68:1542.

Stevens, K. M., and McKenna, J. M., 1957*a*, Very rapid formation of antibody *in vitro*, *Fed. Proc. Fed. Amer. Soc. Exp. Biol.* 16:437.

Stevens, K. M., and McKenna, J. M., 1957*b*, Antibody production in a completely *in vitro* system, *Nature London* 179:870.

Strangeways, T. S. P., and Fell, H. B., 1926, Experimental studies of the differentiation of embryonic tissues growing *in vivo* and *in vitro*, *Proc. R. Soc. London Ser. B* 99:340.

Svejcar, J., Pekarek, J., and Johanovsky, J., 1968, Studies on production of biologically active substances which inhibit cell migration in supernatants and extracts of hypersensitive lymphoid cells incubated with specific antigen *in vitro*, *Immunology* 15:1.

Tao, T. W., 1964, Phytohemagglutinin elicitation of specific anamnestic immune response *in vitro*, *Science* 146:247.

Tao, T. W., and Uhr, J. W., 1966, Primary-type antibody response *in vitro*, *Science* 151:1096.

Terent'eva, E. I., 1955, Experimental cytological analysis of hematopoiesis, Doctoral dissertation, Moscow.

Terskikh, V. V., and Kondratenko, V. G., 1962, Development of primary bone marrow cultures, *Zh. Obshch. Biol.*, No. 2, 153.

Testa, N. G., and Lajtha, L. G., 1972, Some factors affecting survival of CFU and CFUC in culture, in: *In Vitro Culture of Hemopoietic Cells*, p. 102, Publication of the Radiobiological Institute TNO, Rijswijk.

Theis, G. A., and Thorbecke, G. J., 1970, The proliferative and anamnestic antibody response of rabbit lymphoid cells *in vitro*, *J. Exp. Med.* 131:970.

Thiery, J. P., 1960, Microcinematographic contributions to the study of plasma cells, in: *Ciba Found. Symp. Cellular Aspects of Immunity*, London, p. 59.

Thiery, J. P., and Bessis, M., 1956*a*, La genése des plaquettes sanguines àpartir des mégacaryocytes observée sur la cellule vivante, *C. R. Acad. Sci. Paris* 242:290.

Thiery, J. P., and Bessis, M., 1956*b*, Méchanisme de la plaquettogenènes: Étude *in vitro* par la microcinématographie, *Rev. Hématol.* 11:162.

Thomas, M. C., 1956, The growth and development of human leucocytes *in vitro* with particular reference to leukamic cells, M.Sc. thesis, Cambridge University.

Thor, D. E., Jureziz, R. E., Veach, S. R., et al., 1968, Cell migration inhibition factor released by antigen from human peripheral lymphocytes, Nature London 219:755.

Till, J. E., and McCulloch, E. A., 1961, A direct measurement of the radiation sensitivity of mouse bone marrow cells, Radiat. Res. 14:213.

Timofeevskii, A. D., and Benevolenskaya, S. V., 1926, Explantationsversuche von weissen Blutkörperchen mit Tuberkelbazillen, Arch. Exp. Zellforsch. 2:31.

Timofeevskii, A. D., and Benevolenskaya, S. V., 1927, Tuberculous inoculation in the cultures of leucocytes of the human blood, Arch. Exp. Zellforsch. 4:64.

Trentin, J., Wo, N., Cheng, V., et al., 1967, Antibody production by mice repopulated with limited numbers of clones of lymphoid cell precursors, J. Immunol. 98:1326.

Trowell, O. A., 1954, A modified technique for organ culture in vitro, Exp. Cell Res. 6:246.

Trowell, O. A., 1955, The culture of lymph nodes in synthetic media, Exp. Cell Res. 9:258.

Trowell, O. A., 1959, The culture of mature organs in a synthetic medium, Exp. Cell Res. 16:118.

Trowell, O. A., 1965, Lymphocytes, in: Cells and Tissues in Culture, Vol. 2, (E. N. Willmer, ed.), p. 96, Academic Press, London.

Umiel, T., Globerson, A., and Auerbach, R., 1968, Role of the thymus in the development of immunocompetence of embryonic liver cells in vitro, Proc. Soc. Exp. Biol. N. Y. 129:598.

Unanue, E. R., 1972, The regulatory role of macrophages in antigenic stimulation, Adv. Immunol. 15:95.

Van Furth, P., and Cohn, Z. A., 1968, The origin and kinetics of mononuclear phagocytes, J. Exp. Med. 128:415.

Van Herwerden, M. A., 1923, Cultures de moelle osseuse en dehors de l'organisme, Arch. Neerl. Physiol. 8:592.

Vasiliu, T., and Stoica, V., 1929, Culture in vitro du sang du lapin, C. R. Soc. Biol. Paris 100:691.

Vereshchinskii, A., 1924, Über die freien Zellen der serösen Exsudate, ihren Uhrsprung, ihre Genetischen Wechselbeziehungen und ihre prospektiven Potenzen, Haematologica 5:41.

Virolainen, M., 1968, Hematopoietic origin of macrophages as studied by chromosome markers in mice, J. Exp. Med. 172:943.

Virolainen, M., and Defendi, V., 1968, Ability of hematopoietic spleen colonies to form macrophages in vitro, Nature London 217:1069.

Volkman, A., 1955, The origin and turnover of mononuclear cells in peritoneal exudates in rats, J. Exp. Med. 124:241.

Wagner, H., and Feldman, M., 1972, Cell-mediated immune response in vitro. I. A new in vitro system for the generation of cell-mediated cytotoxic activity, Cell. Immunol. 3:405.

Weber, W. T., 1966, Difference between medullary and cortical thymic lymphocytes of the pig in their response to phytohemagglutinin, J. Cell Physiol. 68:117.

Wegmann, T. G., Hellström, J., and Hellström, K. E., 1971, Immunological tolerance: "Forbidden clones" allowed in tetraparental mice, Proc. Natl. Acad. Sci. U.S.A. 68:1644.

Williams, T. W., and Granger, G. A., 1969, Lymphocyte in vitro cytotoxicity: Correlation of derepression with release of lymphotoxin from human lymphocytes, J. Immunol. 103:170.

Wilson, D. W., 1963, The reaction of immunologically activated lymphoid cells against homologous target tissue cells in vitro, J. Cell. Comp. Physiol. 62:273.

Wilson, D. W., 1965a, Quantitative studies on the behavior of sensitized lymphocytes in vitro. I. Relationship of the degree of destruction of homologous target cells to the number of lymphocytes and to the time of contact in culture and consideration of the effects of isoimmune serum, *J. Exp. Med.* **122**:143.

Wilson, D. W., 1965*b*, Quantitative studies on the behavior of sensitized lymphocytes *in vitro*. II. Inhibitory influence of the immune suppressor, imuran, on the destructive reaction of sensitized lymphoid cells against homologous target cells, *J. Exp. Med.* **122**:167.

Wilson, D. W., 1967, Lymphocytes as mediators of cellular immunity: Destruction of homologous target cells in culture, *Transplantation* **5**:986.

Wilson, F. D., O'Grady, L., McNeill, C., and Munn, S., 1974, The formation of bone marrow derived fibroblastic plaques *in vitro:* Preliminary results contrasting these populations to CFU-C, *Exp. Hematol. Copenhagen* **2**:318.

Winfield, J. B., 1971, Role of phagocytosis in macrophage function in an *in vitro* primary immune system, *Cell. Immunol.* **2**:690.

Winkelstein, A., and Craddock, C. G., 1965, Comparative response of normal human thymus and lymph node cells to phytohemagglutinin in culture, *Blood* **26**:876.

Wolf, B., and Stavitsky, A. B., 1958, *In vitro* production of diphtheria antitoxin by tissues of immunized animals. II. Development of a synthetic medium which promotes antibody synthesis and the incorporation of radioactive precursor, *J. Immunol.* **81**:404.

Wolf, N. S., and Trentin, J. J., 1968, Hemopoietic colony studies. V. Effect of hemopoietic organ stroma on differentiation of pluripotent stem cells, *J. Exp. Med.* **127**:205.

Wolff, E. 1960, Sur une nouvelle modalité de la culture organotypique, *C. R. Acad. Sci. Paris* **250**:3881.

Wolff, E., and Haffen, K., 1952, Sur une methode de culture d'organes embryonnaires *in vitro*, *Tex. Rep. Biol. Med.* **10**:463.

Wolstencroft, R. A., and Dumonde, D. C., 1970, *In vitro* studies of cell-mediated immunity, *Immunology* **18**:599.

Woodliff, H. J., 1958, Glass substrate cultures of human blood and bone marrow cells, *Exp. Cell Res.* **14**:368.

Woodliff, H. J., 1964, *Blood and Bone Marrow*, Eyre & Stottiswode, London.

Worton, R. G., McCulloch, E. A., and Till, J. E., 1969, Physical separation of hemopoietic stem cells from cells forming colonies in culture, *J. Cell. Physiol.* **74**:171.

Wu, A. M., Siminovitch, C., Till, J. E., and McCulloch, E. A., 1968, Evidence for a relationship between mouse hemopoietic stem cells and cells forming colonies in culture, *Proc. Natl. Acad. Sci. U.S.A.* **58**:1209.

Yoffey, J. M., 1973, Stem cell role of the lymphocyte-transitional cell (LT) compartment, in: *Haemopoietic Stem Cells, Ciba Found. Symp.* **13**(*New Ser.*):5.